Love's Story (

Why We Are Here

And what we can do about it

Francis O'Neill

A *Making Sense of It* book

Love's Story of Why We Are Here
And what we can do about it

ISBN: 9780993462641 (Paperback)

A catalogue record for this book is available from the British Library (UK)

March 2020. A new jacket was added. A new image, Copernicus Heliocentric Model was added. Also minor revision of content was made – including changing name "Souldiers of Love" to "Soldiers of Love" throughout.

Editor: **Annie Locke**

Cover image by: **Snapwire** (modified)

Printed by: **Kindle Direct Publishing**

Making Sense of It series of books
This series was begun with *Life and Death: Making Sense of It*. These are Mind Body & Spirit books linked together by the common thread of aiding better understanding of spirituality and spiritual health.

Love's Story book is a companion of the *Life and Death: Making Sense of It*. book.

Other titles in the Making Sense of It series include: **Life and Death: Making Sense of It | Steps to Health, Wealth & Inner Peace | Christmas Past**. For availability and updates on these titles please visit the publisher website.

Some Inspiration Publications
SomeInspiration.com
Cotswolds, UK

Please leave a review of this book

As an author I regard the feedback of my readers highly. If in the course of reading this book you have decided that you enjoy it, would you please consider leaving a thoughtful review of the book; on the bookstore where you purchased it. I'd love to hear how it helped you or impacted on you. I'd really appreciate hearing your thoughts on this book. Thank you.

Contents

Preface

Squaring the circle

If you have ever wondered how we might square orthodox science, quantum theory and the theory of evolution, with spirituality; this book offers you a possible solution. What is presented here, I believe, goes some way to square that circle.

In my opinion, there can be no true spiritual understanding of life without including the story of evolution, as well as all the runners and riders that make up this amazing world we live on.

Introduction

What is the meaning of human life, or indeed all of life? Why are we all here? Is there a purpose to our existence, other than what we make of it ourselves? And when we use the terms "soul," or "spirit," to describe the source of our being, are we really only talking about our human selves or possibly something bigger, something that includes all of life in its varied forms and expressions?

I suggest that these are questions which we are all faced with finding answers to. It surely goes with the territory of being human and not least due to being aware of our own mortality. That said, for the most part, these are questions that a great many of us, with our busy lives, tend to overlook. We choose to ignore them or leave to others to answer them for us. But are we short-changing ourselves when we do?

Imagine for a moment that you are a soil scientist, and you have arrived on Mars. You are going to be spending an earth year there studying the soils in a given area of the planet. This is with the view of eventually growing plants that could in the long term contribute towards changing everything regarding planetary habitation and atmosphere. What a wonderful, exciting and valuable adventure you are on. Before arriving you would, of course, have researched all that we know already regarding Mars, and our hopes for producing life there. You would have a pretty clear agenda of what you are embarking on. You would be part of a large international team doing what we humans do best. And you would have answers, albeit probably non-spiritual answers, to some of the questions I posed above – certainly with regard to the vision behind the trip and immediate intentions at least. You may not be contemplating any kind of bigger agenda for Mars and human life as yet, other than what's in front of you, but as an outcome, you would no doubt be encouraged to be looking at life more philosophically, if for no other reason than because of what you are up to.

Meanwhile, back on Earth, I believe that we have arrived on this planet with an agenda. The difference being of course is that we arrive as babies. As babies we are not very good at recalling or setting agendas without help. To fill that gap we come to rely on our parents, our culture, our education and peers etc., to help give form and substance to our agenda for us. And, rather than the quick adjustment required of our soil scientist, for their stay on Mars, here we literally grow into our environment. It becomes what is familiar to us, and it can, depending on how we have been taught to view it, and our world for that matter, also become invisible to us. What I mean is that familiarity tends to breed invisibility. We no longer see what is around us with awe and regard, as we would do on arriving on Mars. We lose touch with the adventure we are on; unless we somehow reawaken that awe within ourselves – or it is helped by other means to be awoken.

Imagine, for instance, those noisy sparrows, that we hardly pay attention to, in our garden each day, now flying past us on Mars. Suddenly we'd be dealing with the unfamiliar, things out of context. We'd probably stop and stare, be in awe of what we had just witnessed, or even spooked by it. If we can disregard the "familiar" for a moment, we will notice we are in this phenomenal situation that we call "Life," and also fortunate to be conscious observers of it. It is truly remarkable that we are participating in it as observers and creators. So yes, to return to my earlier comments, there is, I believe, an obligation upon each of us to explore, understand and attempt answers for the bigger questions life throws at us. We need to make sense of it all to the best of our own ability, before the opportunity passes us by.

I am, of course, intending to go a lot further than Mars in this book, and to also focus on spiritual answers to the questions I'm posing.

At the back of my mind...

In my first book, *Life and Death: Making Sense of It*, I discuss what I see as our spiritual condition, from evidential, philosophical and esoteric perspectives. Primarily I do this through raising a series of ten big questions pertaining to our lives and spirituality. I spend much of the rest of the book providing grist-to-the-mill possible answers – and/or evidence to aid our finding answers to these questions, for ourselves.

That exploration, by the way, involved researching a number of topics associated with the paranormal, near-death experience, past lives, mediumship, the soul and what we call the Other Side. And, not least, it was an exploration of our development as human beings, from our primitive hominid origins. Overall, what I was seeking to address, with the book, is our needing to become familiar with what, I argue, is our true reality, our spirituality, the spiritual world we exist in.

Primarily the book's concern is with getting to an understanding of where we are, and where we are heading as individuals. This I see as being towards consciousness, enlightenment; to freeing ourselves from, what eastern religions describe as the wheel of samsara – and away from our attachment to this transitory situation we exist in.

That given, at the back of my mind, while writing the book, and indeed as a hint running through it, was the idea that there is another perspective to share on this remarkable existence we all lead – however which way it all works out for each of us. This perspective is not just about us, human beings, but about the bigger "us," that is the flora and fauna, the great collective on our planet; and also elsewhere, beyond the Earth for that matter.

Here, in *Love's Story*, I address this other viewpoint. I have wanted to bring this together for quite a while and I am excited that I've actually now got it down onto paper – or "e-paper" if that is how you are reading it.

This book offers and tackles a bigger reason, a bigger agenda for why we are all here. Love's Story tackles what the purpose of our learning and unfoldment is for. It offers a hypothesis for our collective existence, and makes some suggestions which I hope you will find thought-provoking and helpful. When thought through I believe that, for all their complexity, you'll find the hypothesis and suggestions I offer are founded in common sense, with a good dash of logic.

I'm of the view that deep down we know we are not just here for ourselves; to unravel and make sense of our own destiny. We are here because we are a part of something bigger – much, much bigger.

A brief outline

To help you get a sense of the direction of the book here's a brief outline of what it has in store. As is evident from the title, it seeks to answer one big question, "Why are we here?" And, in context, provide advice on "what we can do about it."

The answer to the *Why...* question is best summarised in Love's Agenda, but all chapters are intended to make a contribution to exploring the question and providing the answer – or are designed to clarify parts of it. The book is divided into two parts:

Part One: This part explores Gaia; an issue with evolution; quantum theory; the concept of the soul; and life elsewhere other than here. In this part my intention is to flag up and discuss some of the main contributing topics that both interest me and have had a direct influence on the direction and development of the hypothesis behind what I've called Love's Agenda. It may help to keep this in mind then as you read through these chapters.

Part Two: Explores a model of awakening; the principles operating behind life; why we are here; and what we can do about it. It also offers an illustrative story to help explain the "Love's Agenda." This part presents and discusses the nuts and bolts of the hypothesis.

Epilogue: The Epilogue revisits the UFO phenomenon, discussed in Part One, and considers some of the implications anticipated by it. It also provides a number of Web links for global organisations with serious interest in UFOs and extraterrestrial life.

A comment on "awakening" and "consciousness"

These terms appear frequently in the book and are explored more thoroughly in Part Two. Even so, some explanation of what they mean, to me, may help from the outset, to aid your enjoyment and grasp of the book...

In simple terms, awakening is becoming aware, at increasingly higher levels, of our connection to each other, to the planet and to the universe and beyond. We see more clearly how what we do has an effect on ourselves, on those around us, on our environment and so on, because we become more aware (conscious) that everything is interdependent. Everything we do spreads out and has implications for other things, sometimes far beyond us – and in ways that we couldn't possibly envisage or imagine.

In this book I'm using the terms awakening and consciousness in context with "soul," as being interchangeable with the term soul and spiritual endeavour.

Soul, I believe, will be found to reside in the primordial and primitive up to the highest levels of pure love and light – and I also believe, consciousness spans likewise. It is in process of awakening through levels, from the dormant and unconscious, to the instinctual, to the semiconscious, to the self-consciousness, and on to the spiritual.

Mostly it is of a work in progress, on an organic spectrum of seeding, growing, opening and flowering.

One last thought... on frogs and princes

To plant the seed of a thought with you, as a taste of what I'll be discussing, here's a question for you, a hint at where we will be going with this:

Could a frog become a prince, or a princess for that matter?

If this sounds like a ridiculous or cryptic question, it is not intended to be so. Let me put it another way: If we can accept a frog is a soul in, what I'd describe as a learning and experiencing "situation," that is the situation of being a frog, could not that same soul, further down the line, become human, and a prince or princess to boot? Think on this matter as we journey together through the following. What I'm driving at will all become much clearer in the following pages...

Of course, bring that pinch of salt with you too, in case you think you'll need it. Some concepts are hard to swallow, even when they may contain more than a grain of truth. I trust, however which way you find my arguments, observations and suggestions, this book will spark your imagination and passion to explore. That it will galvanise you into to considering new possibilities, a more profound take on life, and a thirst to look in on, and answer for yourself, some of the bigger questions that life poses.

Let me also add here, in case you don't know much about me, that I am not a scientist, and certainly not with regards to what I want to lay out before you in this book. Some of it, at least, draws upon science. I warn you, there may be gaps in my evidence. I view this work as getting across ideas, philosophical speculation, a hypothesis on life and soul. Besides, a lot of it can't be easily evidenced one way or the other. It is belief. But that is not to say it is beyond the bounds of possibility, beyond reason, or beyond being true.

Enjoy the book.

Francis ☺'Neill
Summer 2018, Cotswolds, UK

Part One

In Part One my intention is to flag up and discuss some of the main "lines of enquiry," that both interest me and have had a direct influence on the direction and development of the Love's Agenda hypothesis.

While linked together, the topics here form a part of the matrix of the hypothesis. They can also be read out of sequence, as chapters independent of each other. I would encourage you to read all though, before reading Part Two.

Chapter 1

The Living Earth (Gaia)

L et's begin this exploration by considering our Mother Earth. As I see it, an important influence on my wanting to write this book lies with our planet...

I wonder, could our planet be acting as one system with a remit to sustain life? Could it be alive, even showing signs of consciousness?

> If there is one signal that will raise the collective hackles (and the guard) of professional science, it is any hint of intentionality. The great commandment of the guild is: "Thou shalt not endow nature with goals, purposes, sentience, values," except where human beings are concerned.
>
> John Newport[1]

Gaia arising

The notion of "Gaia" has been around for a long time. In Greek mythology Gaia is the ancestral mother of all life. She was believed to be alive – of course that stands to reason. She is the original pagan earth mother goddess. Her origins may stretch as far back as into Mesolithic (middle stone age) times, or further. I'd suggest (by their nature) to the time of the "Venus figurines,"[2] at least. The later burial practices of the Neolithic and Bronze Age periods suggest to me an established emotional bond with the earth. With those, that spent their lives in long houses, being interned in long barrows; and likewise those that spent their lives in round houses being interned in round barrows. To live and be reborn in the underworld of the earth – perhaps before re-emerging into physical life. Particularly the Bronze Age burials, being placed in circular mounds, with bodies in the foetal position, are strikingly reminiscent of a return to the pregnant womb of the earth mother.

It isn't difficult to imagine the anthropomorphic relationship our early ancestors are likely to have had with their environment and with nature as their mother. This was a mother that could give on one hand; by providing resources for all their needs, their shelter, clothing, their security, their food, and their children. And, on the other, it could take away, by unwelcome seasonal changes, by weather, and by being awesome, violent and destructive in mood. It could provide traps and surprises, the unexpected predator or pitfalls. It could bring illness and death. Nature is moody and can be light or dark, day or night in appearance. This is a great mother in need of being appreciated, appeased and befriended too. You avoid doing anything to upset her. Rather you respect her and work with her – never against her. If you do work against her, you may pay dearly.

Of course, as we "grew up," supposedly, and began to look at life differently, a lot of us, in the West particularly, put aside many of the "superstitions" of our past. Nature was no longer allowed to be the "all powerful" reason for everything. We learnt instead that we had to control her. The idea of Gaia, the earth mother, became somewhat lost into our misty past, a myth no more, no less.

> The "control of nature" is a phrase conceived in arrogance, born of the Neanderthal age of biology and philosophy, when it was supposed that nature exists for the convenience of man.
>
> Rachel Carson (marine biologist)[3]

Gaia, the mother, the feminine, went underground, let's say, but only to a point. It was retained, indeed central to many beliefs and religions that could be described as environmentally oriented. These include Paganism and Taoism for example. More recently the growths of new age, and holistic beliefs and philosophy, together with the rise of feminism, have all surely contributed towards clearing a way for Gaia's return. Aside from belief, a number of our sciences, led by pioneers like the conservationist, Rachel Carson, began to argue for the simple common sense of working with, and within, the balance and boundaries of nature. We are, after all, living on this one planet, at this time at least, and only on the obvious condition that our planet remains healthy and can support us.

One might suggest this up-swell of holistic interest, and the emergence of sciences concerned with the environment, set the stage for the return of Gaia. Even so, given its primitive roots and pagan pedigree, it was I imagine, a fairly tough call proposing a hypothesis that directly invoked her. But this is what James Lovelock did back in the 1960s. After all here was a highly-respected scientist and environmentalist (working for NASA[4] no less) putting forward what he called the "Gaia hypothesis" (later, when watered down, to become the "Gaia theory"). He bravely, some would say stupidly, put his head above the parapet, when he argued that the biosphere of our planet is a "*self-regulating entity with the capacity to keep our planet healthy by controlling the chemical and physical environment.*"[5]

If you think about it, this was a pretty radical statement, particularly for a scientist to make. Nay, I'd go so far as to say, it was awesome. And he was only scratching the surface of his view by making it. To go a little deeper, Lovelock's Gaia hypothesis posited that:

> The Earth is a self-regulating complex system involving the biosphere, the atmosphere, the hydrosphere [waters] and the pedosphere [soils], tightly coupled as an evolving system. The theory sustains that this system as a whole, called Gaia, seeks a physical and chemical environment optimal for contemporary life.[6]

The biogeochemist, Axel Kleidon, explained this hypothesis further; that Gaia:

> Evolves through a cybernetic feedback system operated unconsciously by the biota [flora and fauna], leading to broad stabilization of the conditions of habitability in a full homeostasis [able to remain constant in spite of varying external conditions]. Many processes in the Earth's surface essential for the conditions of life depend on the interaction of living forms, especially micro-organisms, with inorganic elements. These processes establish a global control system that regulates Earth's surface temperature, atmosphere composition and ocean salinity, powered by the global thermodynamic disequilibrium state of the Earth system.[7]

In simple terms the hypothesis argued that the Earth, in cooperation with the organic and inorganic, is operating a balancing act to maintain and sustain life. But not only that, as life changes and develops (and, in my view, becomes more "refined" – vis-à-vis contemporary life against the flora and fauna of antiquity) so, it is, or was claimed, the Earth also seeks to support this development.

In some respects the hypothesis is analogous to how the thermostat in our homes, or in our bodies, work. We may set the thermostat in our home to a comfortable living temperature. When the ambient temperature falls below this setting, the heating system comes on. When the temperature in the house reaches the desired target again, the heating switches itself off.

Something more complicated, but with similar effect, goes on in our bodies. Given that we are healthy; every one of us operates at a comfortable 37.0 degrees Celsius (98.6 degrees Fahrenheit) now and almost always.[8] If our body temperature deviates outside of what is a narrow range of this temperature, we could die. The human body has a number of self-regulatory, or homeostatic, mechanisms to maintain this balance. In the Gaia hypothesis the Earth is argued to do similar.

As Lovelock pointed out, the idea behind the hypothesis was not entirely without precedent in other sciences:

> The existence of a planetary homeostasis influenced by living forms had been observed previously in the field of biogeochemistry, and it is being investigated also in other fields like Earth system science. [9]

But, thus far, if you read between the lines, on these commentaries they are pulling their punches on the deeper implications of Lovelock's Gaia hypothesis. This, for me, is where it gets really interesting – but was, as it transpired, an embarrassment for Lovelock. He took the hypothesis further – further than Kleidon might suggest. He was making a new departure by clarifying that:

> Gaia is the Earth seen as a single physiological system, an entity that is alive at least to the extent that, like other living organisms, its chemistry and temperature are self-regulated at a state favourable for life ... It is concerned with the working of the whole system not with the separated parts of a planet divided arbitrarily into the biosphere, the atmosphere, the lithosphere and the hydrosphere.[10]

And:

> The originality of the Gaia theory relies on the assessment that such homeostatic balance is actively pursued with the goal of keeping the optimal conditions for life, even when terrestrial or external events menace them.[11]

So take a deep breath... What one can deduce from these comments is that if, "*homeostatic balance is actively pursued with the goal of keeping the optimal conditions for life*," then Lovelock was both suggesting the Earth is "alive" and that it is also proactive in its endeavours. To be clear, in its wildest form, this was implying our planet is a living being, that is maintaining an intentional state of balance; and it is also, to some degree thereby, aware, awake to its role. In effect, keeping an eye on us lifeforms and having our best interests at heart!? Could this be possible, that we are living on a living planet, the living Earth, and thereby with a level of consciousness?

Of course the scientific community would have something to say about this, and you can probably anticipate their response... But, before we look at that, let us first look at what was going on to support erecting such a hypothesis in the first place.

Global conditions supporting Gaia

The development of a global and "living" hypothesis to account for what is going on, on the planet, is not that surprising. Much of what is observed supports the notion of some kind of intentional steady-state system running. To look for evidence for such a system, one only needs to observe elements of the bigger picture.

For example, regulation of the salinity in the oceans: Ocean salinity is understood to have been at a constant of about 3.4% for a very long period of time. This stability is important for life as *"most cells require a rather constant salinity and do not generally tolerate values above 5%."*[12] Before the Gaia hypothesis was put forward this constancy of ocean salinity was an unsolved enigma. It wasn't understood as to why river salts hadn't raised the ocean salinity to a much higher level than observed.

Another example, the atmosphere: There is an observable regulation of oxygen in the atmosphere. This has remained *"fairly constant providing the ideal conditions for contemporary life."* While at the same time, *"all the atmospheric gases other than noble gases [i.e., helium, neon, argon, krypton, xenon and radon] present in the atmosphere are either made by organisms or processed by them."*[13] Invoking a whole system hypothesis helped to explain how the Earth's atmospheric composition has been kept at a dynamically steady state for the benefit of life. Arguably this state could also be explained as entirely due to the presence of life balancing itself, rather than relying on the planet to do it.

And yet another example is regulation of the global surface temperature: It is argued that since life began on the Earth, the energy provided by the Sun has increased by some twenty five to thirty percent, while yet the surface temperature of the planet has remained tolerable for life and habitation. In context, Lovelock also hypothesized that *"methanogens produced elevated levels of methane in the early atmosphere, giving a view similar to that found in petrochemical smog..."* This, he suggests, *"tended to screen out ultraviolet until the formation of the ozone screen, maintaining a degree of homeostasis.*"[14] Importantly, this balancing act has been going on for a very long time – up to 4 billion years – and, notwithstanding human activity threatening this balance, as it is at this time, so it is expected to continue.

Criticism of the hypothesis – and its division into two

Well, as you no doubt guessed, the hypothesis, as was originally proposed (by Lovelock), did not go down well with the broad consensus of the scientific community. It was ridiculed from the bat. The notion that the Earth could somehow be self-regulating and alive was, and still is, seen as wacky, bordering on mysticism and pseudoscience.

In fairness either way, it was, in any case, always going to be difficult testing the bigger claims of such a hypothesis. It had to be broken down into something more manageable and digestible. One of its biggest critics was the Earth and Planetary Science professor, James Kirchner.[15] In his arguments for and against the hypothesis, he divided it into two strands to better aid description and research. These are what he called, the *weak* forms of Gaia and the *strong* forms of Gaia. The weak Gaia asserts that life:

> … collectively has a significant effect on Earth's environment … and that therefore the evolution of life and the evolution of its environment are intertwined, with each affecting the other…[16]

He added that, "*abundant evidence supports these weak forms of Gaia, and that they are part of a venerable intellectual tradition.*" And in that comment he was really making two points: The evidence for "weak" Gaia being already there, and therefore a valuable theory. But he was also implying that there was nothing really new in it, as a theory. Rather it was giving a new name to what was already known and practiced.

The "strong" Gaia, on the other hand, was the sticking point. This was the part that suggested that something bigger was going on that tended to make the environment stable, and enabled the flourishing of all life.

> By contrast, the strongest forms of Gaia depart from this tradition, claiming that the biosphere can be modelled as a single giant organism ... or that life optimizes the physical and chemical environment to best meet the biosphere's needs...[17]

On this he added that, "*the strong forms of Gaia may be useful as metaphors but are unfalsifiable, and therefore misleading, as hypotheses.*" In other words, any claims for the Earth being alive are untestable, and therefore not scientific.

More in-keeping with the general criticism, the CLAW hypothesis,[18] which was inspired by the Gaia hypothesis, was developed to help explain, from a weak form standpoint, how the whole system could work:

> The CLAW hypothesis ... proposes a feedback loop that operates between ocean ecosystems and the Earth's climate. The hypothesis specifically proposes that particular phytoplankton [inhabiting the ocean surface and able to regulate their own population] that produce dimethyl-sulphide are responsive to variations in climate forcing, and that these responses lead to a negative feedback loop that acts to stabilise the temperature of the Earth's atmosphere.[19]

So, the Sun heats the Earth, which warms up the oceans, which encourages phytoplankton growth, which, with the plankton then also dying, causes dimethyl-sulphide (DMS) liquid emissions (smelly flammable liquid from rotting vegetation) which enhances sulphur dioxide in the atmosphere, which in turn encourages more cloud formation, to cover the Earth, which protects it from the Sun, and in turn cools the atmosphere down – and so on. Quite a system... But this was a watered down Gaia – not even retaining its name.

Margulis enters the fray

In the noise of protest, from scientific quarters, against the hypothesis, the value of what Kirchner, and others, were proposing as a viable "theory" (to be found in the weak forms of Gaia), was getting lost and the whole was in jeopardy from its early days. Fortunately Lovelock also had a few outspoken supporters to keep this theory level of Gaia, alive. The most significant was the American microbiologist, Lynn Margulis,[20] who collaborated with Lovelock to help develop it.

Some background on Margulis will be helpful, and, in context with what I'm seeking to get across, has value here. She was both a scientist and an original thinker in her own right. Her work on the origin of cells has helped to transform the study of evolution – well it did so eventually. By the time she was linking up with Lovelock, she had already spent years battling her claims on how evolution works (through her resurrection of the theory of symbiogenesis),[21] with the scientific community, and particularly with those scientists immersed in the zoological tradition. From her perspective they effectively limited evolution, as a study beginning and ending with animals. On which approach she comments, "*All very interesting, but animals are very tardy on the evolutionary scene, and they give us little real insight into the major sources of evolution's creativity.*"[22] She rejected evolution being due entirely to mechanisms of "*mutation, emigration, immigration and the like.*" She instead argued for chemistry and symbiotic mergers as the main driving force in the process.

Her "endosymbiosis theory" proposes that the origins of life stem from bacteria and mergers of bacteria, symbiotic arrangements to form new composite entities, new bacterial cells, and eventually to form larger and varied organisms. This theory has now become mainstream. It was endorsed by none other than Richard Dawkins, who had been in the opposite camp to her. He said:

> I greatly admire Lynn Margulis's sheer courage and stamina in sticking by the endosymbiosis theory, and carrying it through from being an unorthodoxy to an orthodoxy.[23]

Her theory is also an added inspiration, and something of an endorsement, as I read it, for my proposal below. For helping me to make better sense of it and possibly helping to give it legs, so to speak. And she anticipated her theory would get people thinking:

> I think an understanding of the extent to which the evolutionary origin involved symbiogenesis must be acknowledged. Such acknowledgment will lead to new awareness of the physical basis of thought. Thought and behaviour in people are rendered far less mysterious when we realize that choice and sensitivity are already exquisitely developed in the microbial cells that became our ancestors. Even philosophers will be inspired to learn about motility proteins. Scientists and non-scientists will be motivated to learn enough chemistry, microbiology, evolutionary biology, and palaeontology to understand the relevance of these fields to the deep questions they pose.
>
> Lynn Margulis[24]

Margulis first came into contact with Lovelock while looking to understand the metabolism of bacteria and the various gases they produce. She writes, "*Why did every scientist, I asked, believe that atmospheric oxygen was a biological product but the other atmospheric gases — nitrogen, methane, sulphur, and so on — were not?*" She got no answer; but it was suggested that she go talk to Lovelock. It was good advice. Lovelock believes that gases in our atmosphere are all biological. His view is that they are far too abundant to be caused by chemical interaction alone.

Eventually, through her interactions with Lovelock, she got on-board with the Gaia theory. She, like every other scientist, didn't like the idea of the Earth being a living organism, and preferred to see it as an ecosystem. And for her this wasn't about an ecosystem concerned with people per se, but rather it was a biological idea. Nor was it necessarily nice. She writes, *"Those who want Gaia to be an Earth goddess for a cuddly, furry human environment find no solace in it."* She was concerned that reaction to the theory tended to be critical or misunderstood. Either way people tended to judge it, or buy into it, only by misinterpreting it:

> Some critics are worried that the Gaia hypothesis says the environment will respond to any insults done to it and the natural systems will take care of the problems. This, they maintain, gives industries a license to pollute. Yes, Gaia will take care of itself; yes, environmental excesses will be ameliorated, but it's likely that such restoration of the environment will [also] occur in a world devoid of people.[25]

So is Gaia now dead?

More likely she is in hiding again…

As will be clear, from what has been said already, the only way the Gaia theory had gained any real respect (and by "respect" I mean among those few scientists who did support it), was by arguing, from a mechanistic-systems position, for the "weak" forms. There could be no case for a living planet proactively providing balance, stability, homeostatic conditions. There could be no "intentionality" behind it. The notion of a living planet could only result from how things appear to be working, rather than the reality.

In the long run, Gaia has transpired to be a beautiful but unsupportable theory. In more recent times (more noticeably since Margulis's death in 2011) the theory has lost most any favour it had with the scientific community. They are satisfied there was no evidence for making a special case for it. The weak forms of Gaia were already being covered by established sciences so it served no purpose. Back in 2002, Kirchner was already commenting on its demise:

> The most extreme forms of the Gaia hypothesis have generally been abandoned, particularly those that impute a sense of purpose to the global biosphere, and Gaia's proponents have instead searched for mechanisms by which Gaian regulation might evolve by natural selection.[26]

Even Lovelock, according to The Daily Telegraph (of 2016),[27] has renounced his own theory. Gaia was no longer viable. He was never, in any case, really comfortable with the name he chose for it. This was reputedly suggested to him by William Golding (author of Lord of the Flies) on a walk they did together. From early on (as with the CLAW theory) he had also tried, in vain, to disassociate the theory from the mystical and make it more acceptable to science. But few, if any, bought into it. As John Newport explains:

> Challenged or stung by the criticism of his colleagues, Lovelock sought to find some chance-based, nonintentional mechanism to explain the actions of Gaia. ... Lovelock suggested that the name Gaia was chosen merely as a device for communicating with the lay public without need to employ "precise but esoteric language."

However...

Despite these disclaimers, not even Lovelock and Margulis could possibly make sense of their own theory ... without reference to goals, purposes, and intentions. Nothing that is taken to be "self-regulating" can be freed of these attributes. Gaia seen as an active intelligence is emphatically not a metaphor. She is the very substance of the idea.[28]

That said, for the time that the theory was being entertained more thoroughly it caused various sciences, particularly those dealing with environmental matters, to rethink their own concepts and explanations for how things work. Kirchner comments on this:

> One part of Gaia that is clearly fact is the recognition that Earth's organisms have a significant effect on the physical and chemical environment. Biogeochemists have devoted decades of painstaking work to tracing the details of these interactions. Many important chemical constituents of the atmosphere and oceans are either biogenic or biologically controlled, and many important fluxes at the Earth's surface are biologically mediated... This was well understood among biogeochemists well before Gaia, although Gaia's proponents have helped to educate a much wider audience about the pervasive influence of organisms on their environment.[29]

For what it is worth, Gaia has also inspired original research exploring several biologically mediated processes, including (as discussed above) production of dimethyl-sulphide by phytoplankton, and microbial acceleration of mineral weathering. "*This search for Gaian leverage points has proceeded in parallel with a much larger effort by the whole biogeochemical community to trace the mechanisms regulating global geochemical cycles.*"[30] And indeed one could provide a long list of the benefits that the Gaia theory brought to the "rethinking" table. Kirchner again:

> Another well-established fact that is incorporated in Gaia is the notion that Earth's organisms and their environment form a coupled system; the biota

affect their physical and chemical environment, which in turn shapes their further evolution. In this way, Earth's environment and life co-evolve through geologic time...

And getting sciences to think in terms of systems has been a key influence:

A key theoretical element of Gaia is that, as with any complex coupled system, the atmosphere/biosphere system should be expected to exhibit 'emergent' behaviours, that is, ones that could not be predicted from its components alone, considered in isolation from one another. So understanding Earth history and global biogeochemistry requires systems thinking... Gaia's proponents have helped to promote a systems-analytic approach to the global environment, in parallel with a much larger and broader-based effort in the biogeochemical community as a whole.[31]

So is there anything to be retrieved concerning the "strong forms" of the original Gaia hypothesis? Well, while the sciences have taken all that was useful to them from the hypothesis and turned it into a tame systems theory – whilst unceremoniously having dumped the rest – it occurs to me that they would have done that anyway. The more intriguing and fanciful aspect of Gaia was always going to get a rough ride by the sciences. As John Newport's quote pointed out at the top of this chapter, professional science tends to run by the commandment, "Thou shalt not endow nature with goals, purposes, sentience, values." Like anything smacking of the mystical, spiritual or fringe holistic thinking; it gets to be pretty much overlooked or denied, usually without question, in scientific circles. I know it too well from the battering some of my topic interests have received, and unfairly so in my opinion. In this case, they couldn't allow a man, rightfully claiming to be one of their own, making such speculative woo-woo claims.

The only way behaviour changes in science is that certain people die and differently behaving people take their places.

Lynn Margulis[32]

And besides, let us not forget, as Kirchner indicated, with his comments on the stronger forms of Gaia, the concept at that level could not be scientifically tested. It remains therefore an untestable hypothesis. It comes therefore down to *belief*.

As for an old romantic like myself, I have no problem with entertaining the probabilities behind, in this case, ancient belief and healthy speculation. In this meaningful, mythological and symbolical land, Lovelock played a kind of "John the Baptist" figure, when announcing the return of Gaia. She may be back in hiding again now, but I choose to believe in her return. I choose to believe in the possibility of the living Earth concept. Could a planet really be alive and purposely participate in the development of its lifeforms? Yes, I can buy into that notion. As Einstein asserted, all matter is energy. We know so little about the interconnectivity of life, both animate and inanimate, and what mystery is yet to unfold. Let us keep an open mind on such a hypothesis.

Besides, I have an ulterior motive: It supports the ideas I want to get across to you in this book.

Notes & references

[1] Newport, John P. (1997) The New Age Movement and the Biblical Worldview: Conflict and Dialogue. Wm. B. Eerdmans Publishing Company.
[2] "Venus figurines" is a generic term relating to a number of prehistoric statues (carved in soft stone, bone or ivory, or formed of clay and fired) of women sharing common attributes. They stem from the Aurignacian or Gravettian period of the upper Palaeolithic and are found from Western Europe to Siberia.
[3] Carson, R. (1962) Silent Spring. Houghton Mifflin. P297

[4] James Lovelock worked for NASA, developing instruments for the analysis of extraterrestrial atmospheres and planetary surfaces. One of the programs he worked on was the Viking program - which visited Mars, late 1970s, to determine (in part) if Mars ever supported life.

[5] Lovelock, James E (1979) *Gaia: A New Look at Life on Earth.* Oxford University Press.

[6] Lovelock, J. (2009) *The Vanishing Face of Gaia.* Basic Books.

[7] Kleidon, A (Mar 2011) *How does the earth system generate and maintain thermodynamic disequilibrium and what does it imply for the future of the planet?* Article submitted to the Philosophical Transactions of the Royal Society.

[8] Wonderpedia explains why this temperature is important: According to research conducted by scientists in Israel, the reason our body temperature is warmer than the average ambient temperature is to keep fungal infections at bay. For every 1°C degree rise in temperature, the number of fungal species that can thrive on our bodies declines by 6%. It seems that 37°C strikes the perfect balance between being able to prevent infection and the cost of maintaining our body temperature in terms of energy consumption. From http://www.wonderpediamagazine.co.uk/mind-body/why-our-body-temperature-37-degrees-c [Accessed 14/10/2017]

[9] Lovelock (2009) op. cit.

[10] Lovelock, J. (1991) *Healing Gaia: Practical Medicine for the Planet.* Harmony Books: New York

[11] Lovelock (2009) op. cit.

[12] Harvard University (2012) Gaia hypothesis. Gaia-hypothesis-wikipedia (pdf file).

[13] Ibid.

[14] Ibid.

[15] James Kirchner, professor of Earth and Planetary Science at University of California, Berkeley. Kirchner was very active in the evolution of the Gaia hypothesis, though not necessarily a supporter of it.

[16] Kirchner, James W. (2002) *The Gaia Hypothesis: Fact, Theory, and Wishful Thinking.* Climatic Change 52: 391-408. Kluwer Academic Publishers.

[17] Ibid.

[18] The CLAW hypothesis, an acronym of its authors' names: Charlson, Lovelock, Andreae and Warren.

[19] Harvard University (2012) op. cit.

[20] Lynn Margulis (March 5, 1938 – November 22, 2011) was an American biologist and University Professor in the Department of Geosciences at the University of Massachusetts Amherst.

[21] Symbiogenesis, or endosymbiotic theory, is an evolutionary theory of the origin of eukaryotic cells from prokaryotic organisms, first articulated in 1905 and 1910 by the Russian botanist Konstantin Mereschkowski, and advanced and

Love's Story Of Why We Are Here

substantiated with microbiological evidence by Lynn Margulis in 1967. https://en.wikipedia.org/wiki/Symbiogenesis [Accessed 18/11/2017]

[22] Brockman, J (1995) *The Third Culture: Beyond the Scientific Revolution*. Simon & Schuster.

[23] Ibid.

[24] Ibid.

[25] Ibid.

[26] Kirchner, James W. (2002) op. cit.

[27] Blair, T. (October 4, 2016) *Is Gaia Dead?* The Daily Telegraph. http://www.dailytelegraph.com.au/blogs/tim-blair/is-gaia-dead/news-story/61e052eb2f890d05d3d943687e2ae831 [Accessed 31/01/2018]

[28] Newport, John P. (1997) op. cit.

[29] Kirchner, James W. (2002) op. cit.

[30] Kirchner, James W. (2002) op. cit.

[31] Kirchner, James W. (2002) op. cit.

[32] Newport, John P. (1997) op. cit.

Chapter 2

A problem with the Theory of Evolution...

If Gaia can be said to have no "intentionality" behind it, where does that leave life itself?

> The process [of life] is completely random and purposeless, and what looks like goal directed development is actually due to the chance relation between random changes in the organism and the physical characteristics of the ecosystem.
>
> Neville Moray[1]

Is the life we see around us, that we are a big part of, really so pointless, so rudderless, so goal-less, or are we just "not seeing the woods for the trees?" I believe there is another argument to be made on evolution.

It is not so much a problem with "evolution" per se that I'm wishing to address here, but some of the deductions made around it, or from it. I see this is an appropriate point to layout my concerns with evolutionary theory, as it stands, before asking you to head further into a book that relies on a very different interpretation of evolution.

As we understand it through Darwinian (or neo-Darwinian) theory, evolution is responsive and yet blind. It does not provide any explanation for the origins of life nor its direction of travel, nor end objectives. Life just is, in other words. Each species mutates and adapts according to the ecosystem pressures upon it.

In simple terms, that's where the theory of evolution really begins and ends. It's a "now" theory in this sense. We know where species are in their development and adjustment now. Yes, we can know, with hindsight, the route/s a species has taken to get to this "here and now," or, alternatively, how they have become extinct. We may also make predictions on how an organism is likely to mutate in appearance and behaviour going forward, based on models of climate change and changes in the local environment. But that is really it. The bigger questions are left hanging...

Evolution allows us to observe the mutations in creatures and allows us to comment on the "How." How this or that development has taken place, or may take place, in order for a given species to survive and prosper, but it tends to ignore the "Why." Not the little whys, the bigger whys...

The elephant in the backyard

We know a lot about mammals, for instance, not just ourselves (when viewed as animals) but probably all species of mammals that we have observed. They first came on the scene around 210 million years ago – before, the dinosaurs. We know why and how they have offspring, why and how they produce milk. We know, for example, about elephants. We know elephants are related to the extinct mammoth. There are three types that we know of today, the Asian elephant, the African Bush elephant, and the African Forest elephant. They live in female led groups.

They are amongst the largest land animals. The African Bush elephant is indeed the largest land animal – a bull elephant can stand up to four metres tall. We know they are herbivores, and, food-wise they can daily put away around eight percent of their bodyweight. They can also drink up to 60 gallons of water in a day. Elephants are intelligent; we know that. They have the largest land animal brain, inch thick skin, and their trunk can be up to 2.4 metres long. They can weight up to 6 tonnes and have an average life-span of 60 years or more in the wild. The oldest recorded, so far, was *Lin Wang*, an Asian male elephant, who died at 86 years old, at Taipei Zoo, in Taiwan.

It clearly goes without saying, we know a lot about elephants. We can further describe why we think there are different types; that it is due to their development in differing parts of the world – in different environments. It's all brilliant stuff. But now let's consider the elephant and a bigger "why." Why are elephants here? No don't give me the circular answer; because of "evolution." That's not really answering the question that I'm putting forward. It is not because their tusks are at the starting point of a disgusting trade in fine ivory for us humans either. No, I'm asking, why have elephants come into being; for what purpose? What's their end game? I think it fair to say elephants have a huge part to play in helping to maintain a balance in their ecosystem. For example, elephants use their tusks to dig for water that other animals may then benefit from. They also help to spread the seeds of trees and other plants, to help populate the savannahs. But I'm looking for something more meaningful than that. And it's not just about elephants. We needn't stop there. Why do we have birds, bears or bananas? Why life, why us?

Evolution, as it is understood in Darwinian thinking, will only take us so far in answering such questions. Truthfully it can only suggest that the answer to the big "Whys" lies in what is already clear, that is life's response to the environment. While this is obviously valid, and possibly satisfies the academic mind, it conveniently ignores what I would call, *proactive evolution*. Could we all be following pathways that while being meshed into taking from, and contributing to, our environment also might lead us on to something else? This is evolution with intention, direction and purpose – with every lifeform playing a vital part that links in with supply and demand. Granted, there is intelligence operating behind life but the theory of evolution, as it stands, offers no room for intelligent purpose, intentional progress or intelligent design (I think "intelligent design" might be an expletive in some scientific quarters) to enter into the description. It really all comes down to how the dice falls. It is, in other words, seen as neither intentional nor meaningful. We can fairly deduce, from this perspective, that life is therefore unintentional and meaningless – and that is surely the conclusion we are forced to arrive at by this route.

But what if...

But what if the James Lovelock theory of Gaia is really correct, right to its very core? What if the Earth is involved not only with making the environment stable, to enable the flourishing of all life, but is actively pursuing an agenda in support of a proactive evolution? And what if that proactive evolution agenda is in support of the development of something really big?

In other words, what if we are only seeing and measuring one side of the evolutionary equation, an equation that has involved trial and error, modelling and remodelling in search of something? What if the ability (or not) of species to adapt and keep tuned to their environment is therefore only one part of the prism? What if the other side is that the "environment" is dynamic, that it does have intention, a goal in mind? Too bizarre would you say?

But if you stand back and look at it, past our 9 to 5 existence, life is actually bizarre. I mean we are attempting to make sense of this life phenomenon "after-the-fact," after the fact that it already exists, and we are in it. We arrived after it all got started. We are a part of it and therefore obliged to make sense of it – at least to attempt to do so.

It started with a kiss...

What if all this life stuff around us stems from the success of some bacteria invading other bacteria, and later, not only deciding that they could work together but that they actually became as one unit? Could it be possible that the resulting micro-unit of life also had big dreams and went on to form life as we know it? We are talking of a relationship that probably began between three to four billion years ago.

Let me bring back Lynn Margulis here. In her research, she reintroduced a long overlooked idea that the organelles (tiny cellular structures that perform specific functions as the nucleus within cells) within eukaryotes (living cells that form the organisms of life) were once upon a time free-living bacteria. These bacteria had then invaded other bacterial forms. At first they are thought to have become parasites, living off their bacterial host. Later they began forming a symbiotic relationship with their host, providing services in return for a protective environment. Even later they became fully integrated into their host's biological make-up; becoming the organelles.[2] A new force, to be reckoned with, was then being born – the "CELLS." From these cells, we can say, the *myriad creatures have appeared.*

> Therefore I should infer from analogy that probably all the organic beings which have ever lived on this earth have descended from some one primordial form, into which life was first breathed.
>
> Charles Darwin[3]

But before all that... Just to be clear, the bacterium that became the organelle was not the first, "primordial form" of life that Darwin is referring to here. What that primordial form is, is referred to today as the "last universal common ancestor," or LUCA for short. By default this predates the three domains of life as we know it, that of bacteria, archaea (cell life having no nucleus) and eukaryote (cell life with a nucleus). It would have been a *"...simpler, more rudimentary entity than the individual ancestors that spawned the three, and their descendants."*[4] The timeline can't be known for sure, as the rocks provide no fossil evidence of it, but it is anticipated that this LUCA must have been kicking around on the planet going back four billion years.

The theory is, obviously, that all life shares a common ancestry with this primitive form of life. And look at the mayhem it has caused. Variety looks to be the spice of life, literally with this organism. It has gone off in a huge number of different directions to arrive at this "here and now."

Environment: visible and invisible

And let me ask you this question at this juncture: Given the environment is argued by the mainstream sciences as the driver for all of this development, has the environment been so remarkably different locally and globally that the forms of life we see around us are (or would be) the expected result? Of course, the lifeforms themselves make up part of that environment. But might there be some other missing link, or driver in the matrix, that is getting overlooked?

I buy into the notion that lifeforms are often in a symbiotic relationship with each other, and with the Earth. It is built on the basis of all working together. I believe they are working towards an end goal. I like, therefore, to consider the idea that there was already a plan in place, a primary objective if you like, for this little LUCA, this rudimentary lifeform, to try every which way to get to its destination.

But is this merely belief on my part? Let us approach this matter from a different perspective: If, as Moray says above, the *"process of life is completely random and purposeless,"* and any goal directed development is *"actually due to the chance relation between random changes in the organism and the physical characteristics of the ecosystem,"* one might ask how come there is so much intelligence, order and determination, in each species, while existing in this randomness? How come there are so many organised species – guesstimates now amounting to a trillion species of plants and animals? How did we end up with so many by chance? Indeed why, if all is "due to the chance relation between random changes," are there not more distortions and abominations in the manifestation of organisms? One would, fairly, expect biological order to be a random thing too.

As you might anticipate, there is also no requirement for the concept of "soul," or indeed anything spiritual in this version of our evolution story. It is an explanation that is obviously materialistic. Although, that said, the "soul" may serve a purpose similar to how Aristotle perceived it, that is as related to an integrated living system rather than something invisible or spiritual. Here's Moray again:

If at all, it is in this [Aristotelian] aspect of a living system that the notion, the philosophical story, of a soul may be useful. But we have to be careful, because very often in everyday stories people take it for granted that souls are parts of a person that are not material, and we have already seen ... there is really no reason for such an idea.[5]

I think, from this comment, Moray betrays the fact that he has never considered turning this concept of "souls are parts of a person" on its head, that is, seeing the body, the whole body at that, being a vehicle of the soul. Relying on "everyday stories" is not the most prescriptive source for describing the soul either. It needs to be studied. For me, his treatment of the matter here doesn't really cut it.

But let's get to another scientific perspective on this, and one of the reasons I wanted to discuss and reconsider the conventional description of evolution here. Through cell research this materialistic approach to the "living system" is undergoing change, and welcome change at that. Consider these comments from Dr Bruce Lipton:

Leading edge contemporary cell research has transcended conventional Newtonian physics and is now soundly based upon a universe created out of energy as defined by quantum physics. This new physics emphasizes **energetics** over materialism, substitutes **holism** for reductionism, and recognizes **uncertainty** in place of determinism. Consequently, we now recognize that [cell] receptors respond to energy signals as well as molecular signals.[6]

The explanation for how organisms behave, and are controlled, was believed to lay entirely in their DNA/genes. The modern approach is to give much more consideration to environmental influences in the development and direction of organisms. This environmental influence is carried through the membrane of cells responding to external or physiological stimulus. Here's Lipton again:

> Cells "read" their environment, assess the information and then select appropriate behavioural programs to maintain their survival. The fact that data is integrated, processed and used to make a calculated behavioural response emphasizes the existence of a "brain" equivalent in the cell. Where is the cell's brain? The answer is to be found in ... its "cell membrane."[7]

But now, what is really exciting to me is the changing explanation for what this "environment," may consist of, that cells are reading. Here Lipton (like Lovelock) touches on the near mystical:

> Conventional medicine has consistently ignored research published in its own main-stream scientific journals, research that clearly reveals the regulatory influence that electromagnetic fields have on cell physiology. Pulsed electromagnetic fields have been shown to regulate virtually every cell function, including DNA synthesis, RNA synthesis, protein synthesis, cell division, cell differentiation, morphogenesis and neuroendocrine regulation. **These findings are relevant for they acknowledge that biological behaviour can be controlled by "invisible" energy forces, which include thought.**[8]

And which comment, to my mind, more than hints at moving the needle a long way towards implicating one particular primary "invisible" energy force that is thinking its way forward. This is the life-force that can also manifest rapidly, and move across lifeforms in its progress towards awakening. I'm thinking of the soul; the key part being left out of the equation in evolutionary theory. No lifeform can exist without it. And does it have an end game? I believe it most certainly does.

Love's Story Of Why We Are Here

Notes & references

[1] Moray, N (2014) Science, Cells and Souls: An Introduction to Human Nature. AuthorHouse UK.

[2] There are other speculative hypotheses as to how the first cell, having a nucleus, came about, but this "syntrophic model" is currently the most accepted. See https://en.wikipedia.org/wiki/Cell_nucleus [Accessed 31/01/2018] for other possibilities.

[3] Darwin, C. (1859), The Origin of Species by Means of Natural Selection, John Murray, p. 490.

[4] Woese, C. R.; Kandler, O.; Wheelis, M. L. (1990). Towards a natural system of organisms: proposal for the domains Archaea, Bacteria, and Eucarya. Proceedings of the National Academy of Sciences. 87. https://en.wikipedia.org/wiki/Last_universal_common_ancestor#cite_note-14 [Accessed 31/01/2018].

[5] Ibid.

[6] Lipton, Dr. B. H. (2000) Biological Consciousness and the New Medicine. http://www.dragonway.co/essays/biological_consciousness.htm [Accessed 04/01/2018]

[7] Ibid.

[8] Ibid.

Chapter 3

Quantum Theory

A snapshot overview

What is quantum theory, and why is it so important in our understanding of the world we live in?

> The idea of matter being frozen light may be difficult to accept for some readers. They will find it even harder to accept the statement ... that matter is spirit grown solid or frozen. Yet the inconceivable wisdom of the Creator is not bound by what men consider reasonable.
>
> Kurt Eggenstein[1]

> If you want to find the secrets of the universe, think in terms of energy, frequency and vibration.
>
> Nikola Tesla[2]

I want to say something about quantum theory here. Bear with me. If this is a topic that you already know a lot about, or, for that matter, if this is a topic that turns you off from the get-go, because you think it is too obscure to grasp, let me tell you that this is not going to be a long dissertation on the matter (no pun intended), but rather a postage stamp size philosophical overview. It allows me to make a comment or two, and an observation germane to the direction of the book. This chapter has bearing on what I'm exploring, and proposing in the second part of the book. The following then is a brief, but I trust succinct and easy to follow overview of the theory.[3] Enough to make it worth your while taking a look at it.

Quantum theory in a nutshell is the theory of atomic phenomena.

It has turned, and is still in the process of turning, the Newtonian worldview upside down. Isaac Newton (and fairly many others, especially Descartes, and Copernicus before him) is attributed with pioneering the development of what became known as the (Western) mechanistic and deterministic view of our world; and worlds beyond for that matter. During the 1680s Newton, drawing on laws, discovered by Kepler, formulated his mathematical laws of universal motion (around gravity) which started the ball rolling for a scientific revolution...

The crux of this is that the Newtonian view established that all knowledge, that is true and reliable, can be known or predicted by the observation and measurement of matter, and/or the forces and laws governing matter. This is knowledge based on the material objective world that, for many of us, is still assumed to be reality indeed the only reality.

From this mechanistic approach we have, of course, come to an understanding of the natural world. It has helped to bring about the industrial revolution and improve the quality of the lives of all who have benefitted from it. Through it we have built our amazing world of structures, machines, tools and instruments. It has allowed us the ability to predict and manipulate the resources that lie beneath our feet, and predict the nature and behaviour of what lies above us, in the movement of planets and stars.

However, equally it can be said that this mechanistic (and essentially masculine) construct on life has, coming forward, proven itself to be risky, out of balance, taking us to the brink of our being more disconnected from each other and closer to destroying ourselves and our planet. This danger continues to exist, for example, in our attempts to control and over-use nature for our own purposes rather than work with it; and in our growing capability for unleashing powers of mass destruction on each other.

World views in collision

It should be noted that the power of orthodox religion, at the time of the scientific revolution, was also a driver for the direction of this developing paradigm – it provided the orthodoxy to push against. This was a triumph of materialism over religion in other words. There was no need for a god in this perspective. Or if there was, it was just to get the machine, the clock, started. Darwin's theory of evolution further provided support for this materialistic direction, and, as we know, provided a further blow to religion.

Fairly, in terms of worldview, we are still living with the aftermath of this power struggle between church and science. But we mustn't overlook a struggle that has been taking place in science itself.

It was inevitable that as science got underway it would explore the micro as well as the macro. And in doing so it assumed to predict behaviours, at the atomic level, to follow the already established mechanistic law/s. After all, atoms are the building blocks of matter, the objective "real world." But would this turn out to be actually the case? Clearly those brave enough to look over that precipice, at atoms, found that it was not at all the same as could be applied to what we see around us. It was operating under different laws:

> The ontology of materialism rested upon the illusion that the kind of existence, the direct "actuality" of the world around us, can be extrapolated into the atomic range... This extrapolation is impossible, however... The naive materialistic way of thinking is an obstacle to understanding the quantum concept of reality.
>
> Werner Heisenberg[4]

When physicists Werner Heisenberg and Erwin Schrödinger began looking at atoms they entered a new world, the land of the unknown. Newtonian predictions soon went out of the window. Instead, what they were finding was staggering and gave rise to the idea that our living in an objective world, of things separate and independent of each other, was no more than an illusion. It is what Buddhism had been telling us all along.

What they found, and what quantum theory demonstrates, is that what we see as matter is actually energy. As you explore deeper into the workings of atoms you see nothing but energy. And that energy is dynamic.

When an atom is split it is said to break into particles of, depending on which atom, electrons, protons, neutrons. These swirl around like in a tornado. The tighter the space they have to move in, the faster they swirl. But these are not particles of matter – as we might want to see or expect to see. There are no physical particles to swirl. They are, what physicists call, "quanta," meaning they are described as wave functions that only allow for the "probability" of particles to appear. There is no certainty that they will. Let's say there is "no certainty" period. They can behave as particles or waves, spontaneously, and as particles they can be in two places at once. We are talking bizarre when compared to the world we know.

An even more bizarre occurrence is that whether, say, an electron behaves as a particle or as a wave, will tend to depend upon whether it is being observed or not – by something that is conscious. What experiments in quantum physics have shown is that:

> ... human consciousness plays a crucial role in the process of observation, and in atomic physics determines to a large extent the properties of the observed phenomena ... In atomic physics the observed phenomena can be understood only as correlations between various processes of observation and measurement, and, the end of this chain of processes lies always in the consciousness of the human observer.
>
> Frijof Capra[5]

In other words, behavioural outcomes will be mostly down to who is watching. It's subjective, not objective. If no one is watching an electron may do one thing. If someone is watching, and/or deliberately measuring it, it will probably do another. It has been said elsewhere that it is, as if, the electron "knows it is being observed," and behaves accordingly.

From Potential to Actual

> ...every atom is continually striving to manifest more life; all are intelligent, and all are seeking to carry out the purpose for which they were created.
>
> Charles F. Haanel[6]

So what, in heaven's name, holds it all together?

If we try to apply quantum theory to the physical world, we are left with a bit of a conundrum. The conundrum lies in what is causing us to see and experience matter (that includes ourselves) as something real and solid? If the building blocks of matter, that is atoms, can be described as not solid, not matter, but as energy, that only have the probability or potential for existence – and seemingly can only be brought into "actuality" by being conscious observation – how does that work for the world around us? Does it only exist because it is being observed, and interacted with, via our being conscious, or is there something else?

There is always the danger, in following this line of enquiry on the theory, that it is human consciousness that is seen to bring everything into actuality. There may be some element of truth in that (from what has been observed in sub-atomic behaviour), but we know though that the world is just as real (although it may be a different world) to an insect or plant as it is to a human. I wouldn't call plants and insects conscious yet, in the manner that we would describe for ourselves. So what else could it be?

It has been muted that it is the great collective of consciousness that forms energy into matter. That this is the cause of the "reality" we see around us. We might then ask would it all disappear if this collective fell asleep. Or would everything change if, while asleep, this collective dreamt of a different reality to wake up to? To my mind, this is a very interesting exercise in lateral thinking, but leans a little too much on the side of fantasy. So what else could it be?

Science doesn't appear to have an answer for this yet – or at least not an answer it can agree on. From my perspective, the clue to what holds things in place is in the frequency and vibration of energy. What holds each in place is what I describe as "soul."[7] Soul provides the organisation, integrity and the boundary to things. And behind this lies, what I perceive as

"Love," something wonderful that holds it all together, and in flux. But it is soul, a spark, a particle, a drop of Love that actualises in what we see around us. As Tesla describes, it is *energy, frequency and vibration* but coupling with soul that creates and holds expression. It is soul bringing things into manifestation, as it stretches to awaken and gain consciousness.

In context I find the following quote, by Max Planck to run on a parallel with my understanding of what is going on:

> As a physicist, that is, a man who had devoted his whole life to a wholly prosaic science, the exploration of matter, no one would surely suspect me of being a fantast. And so, having studied the atom, I am telling you that there is no matter as such! All matter arises and persists only due to a force that causes the atomic particles to vibrate, holding them together in the tiniest of solar systems, the atom. Yet in the whole of the universe there is no force that is either intelligent or eternal, and we must therefore assume that behind this force there is a conscious, intelligent mind or spirit. This is the very origin of all matter.
>
> Max Planck[8]

To be honest, when weighed up, I'm thinking Planck's spiritual conclusion, and my own, as little more than inevitable musings. And while considering this, we surely also get a glimpse, from Planck's comment, into another "probability" heading for "actuality:" What I mean is that because of quantum theory, science is, by its own findings, being forced to make a turn. It is in the process of coming almost full circle. It is heading, not so much back to some kind of mediaeval religious dogma but towards a clearer and yet spiritual construct on life – an holistic-spiritual worldview in other words.

This process will of course take time. The mechanistic paradigm (and those that follow its credo), so successful in its application, is not going to just rollover and transform, without putting up some resistance...

In the next chapter we'll take a closer look at what the soul is understood to be. Note that there are parallels between the mystery of quantum theory and the esoteric understanding of the soul. But that is surely to be expected.

Notes & references

[1] Eggenstein, K. (1984) *Materialistic Science on the Wrong Track.*
http://www.chemtrails-info.de/kee/1/i-mater.htm [Accessed 14/04/2018]

[2] For more content around this popular quote see *Nikola Tesla on the Secrets of the Universe - Energy, Frequency and Vibration.*
https://www.youtube.com/watch?v=rieJef500nU [Accessed 14/04/2018]

[3] Max Plank, Albert Einstein, Niels Bohr, Louis de Broglie, Erwin Schrödinger, Wolfgang Pauli, Werner Heisenberg and Paul Dirac all played a part in developing the quantum theory and its implications for Newtonian physics.

[4] Heisenberg, W. (1962) *Physics and Philosophy.* New York: Harper and Row. p145.

[5] Capra, F. (1983) *The Turning Point.* Flamingo/Fontana Paperbacks. P76.

[6] Haanel, F. C. (2016) *The Master Key System.* Some Inspiration Publications. item 34 p10. First published in 1916 by Psychology Publishing USA. Book can be obtained via Some Inspiration website https://someinspiration.com/books/the-master-key-system/ [Accessed 02/05/2018]

[7] In esoteric description/belief what holds and maintains form is down to devas (or angels). I prefer to keep this simple.

[8] Planck, as cited in Eggenstein, K. (1984) *Materialistic Science on the Wrong Track.*
http://www.chemtrails-info.de/kee/1/i-mater.htm [Accessed 14/04/2018]

Chapter 4

Bless my Soul

What is the soul? What is its nature? Where will it be found? How does it link with Love, and karma for that matter?

This chapter offers an overview of the soul.[1] In context with quantum theory the soul is, I suggest the orchestrator and conductor of energy. In context with Love's Agenda hypothesis it plays a central role – as will become clearer in Part Two. It is therefore imperative to provide a clear description of the soul here. By the way, this is not a book about religion in any conventional or orthodox sense. What I'm calling "soul," throughout this book is my preference. You may use another name; perhaps "spirit" or possibly "higher self." I'm a simple soul at heart and to me the different names or terms we use mostly amount to the same thing – at least in general parlance.

What the soul is

The soul is understood as a life and light source. It vibrates at a level that makes it invisible to our normal physical sensors. It is our being-ness.

To take this description further it will help if we consider the soul as being similar in form to an onion, with a number of layers. These are all vibrating at different frequencies, from slower and coarser layers to faster and finer layers. These

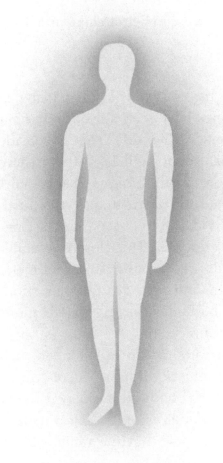

Fig. 4.1 The aura around the human body

layers are identified as vehicles of the soul, with the physical body (at the lowest vibration and frequency) being the outer layer, or vehicle, that we can see. We can say therefore the soul, as a whole, is not entirely invisible but how much we can see of it is relative to our ability, or the technology we have, to view its finer layers.

When we look in the mirror, or at another person, then we are seeing the outer garment of a soul. But of course we'd probably say the physical body also gets in the way of seeing what we might want to better objectify as the soul. Here the aura, around the body, can help to take us a little closer…

The Aura

Everything alive has an *aura*. A great many of us will have heard of, or perhaps have seen, the aura that manifests around the body of a person (Fig. 4.1). This is considered to mostly emanate from the etheric vehicle. If this is new to you I'll be saying a little more about vehicles of the soul below, so bear with me. As an aid to understanding this further, the etheric vehicle may be divided into two parts: The first of these being called the "etheric double." This closely resembles the physical vehicle. It extends out around two centimetres beyond the surface of the skin, and provides the framework on which the material body is built. The second part is more appropriately considered as the aura.

The aura is roughly an egg-shaped field of energies surrounding the etheric double – often more noticeable towards the top of the body. It is claimed the aura can extend out from a few centimetres up to half a metre or so from the physical body – that's within the range of my experience of seeing it too. It serves as the interface between the physical body and the cosmos, *"and all the forces of the universe are reflected on its surface."*[2] The aura is also heavily influenced by the astral or emotional vehicle, and is argued that it is this that actually forms its second part – see Max Heindel's "ovoid cloud" comment below.

The Colours of the Aura

The aura displays in various colours according to the state of physical health, mood and the psyche of the person being observed. Note that it is rare for there to be just one colour – although there will tend to be one dominant colour. The colour, and shade of colour, is also taken into consideration when interpreting a person's state. With regards to the interpretation of colours, you will find a lot of information covering this on the Web. To give you a taste of these; here are a few examples:

Yellow: A predominantly yellow aura suggests an analytical and logical thinker. This colour also has links with the spleen.

Red: A predominantly red aura is linked with the blood. Depending upon shade, it indicates strong enthusiasm for life, and the suggestion of an adventurous person.

Green: A predominantly green aura is associated with the heart. It suggests creativity and a person who seeks perfection. It is also associated with healing – particularly the turquoise shade.

Pink: A pink aura may surround a person who is naturally a very loving and giving person, but who, we might suspect, is also needy of others in their life, in order to be happy and at peace.

Orange: Orange is linked to the organs of reproduction. It suggests a person who is energetic, generous and confident.

Purple: Purple is associated with the glands and nervous system. It suggests intuition and psychic ability.

Blue: Blue is linked to the throat and thyroid. It indicates calmness in thought, with communications being a priority.

You can gather from this that there is a broad range of possible colours that can be seen. These can also include black, white, gold, silver, pastels, muddy colours, and/or rainbows. All are given meaning. As you might imagine, not all auric colours indicate good health and/or a wholesome state of mind. You can probably guess that blacks and shades of grey are not particularly welcome colours: black indicating some kind of negativity, a holding onto something, perhaps a built-up resentment and a possible meanness of spirit. Grey is also negative but more passive than black, suggesting blocked energy and fears, a mistrust of people or life.

Viewing the Aura

It is believed that not all of us can see auras. I'm of the opinion that if we adjust our eyes we can learn to see the aura. It does take practice and patience however.

Do you recall looking at "magic eye" (autostereogram) pictures? They were very popular in magazines and newspapers some years back – certainly in the UK. They can still be readily found on the Web. What these are, are two-dimensional patterns that have a 3D image/s hidden within them. If you look at the image, using normal eyesight, they reveal nothing but an abstract pattern. If you look at them in the right way (I find it helps if I look at it slightly cross-eyed), suddenly the intended 3D image pops out in sharp relief. And indeed, just as suddenly, the rest of the picture also becomes a 3D backdrop for the target image.

What we are doing, when "getting it," is using both our central and peripheral vision together. Seeing an aura requires a similar kind of adjustment in how we use our eyes. Here are some tips you can try to help you develop your ability to see auras:

- o Focus your eyes beyond the person, whose energies or aura you are attempting to see. What you are doing is learning to use both your central and peripheral vision together.
- o Try being slightly cross-eyed if it will help. You are thereby letting the aura reveal itself. It takes a bit of practice to refocus in this way.

- o Get the person you are looking at, to stay still for a period, while you get it. A house plant, I suggest, would be a good alternative to practice on.
- o It also helps, indeed is necessary when learning to focus, if you can view the person or plant with a simple one-colour (e.g. white) or light tone backdrop to help the subtle auric colours then display in relief. I might suggest, for example, setting up a small to medium potted leafy plant against a one-colour wall, then sitting back a few feet, and allowing your eyes to focus on it.

What is happening when we see an aura is that we are observing vibrations, vehicle/s of the soul that our mind, via the brain, interprets into colours. Let's now look more closely at the vehicles I'm describing.

Vehicles of the Soul

It depends on what tradition, and level, you draw from regards how many vehicles the soul is understood to be embedded in.[3] So as not to confuse the matter, here I'm drawing on the most commonly considered vehicles, that are a part of Western Theosophical and Rosicrucian traditions, where the soul is viewed as having four identifiable bodies, or vehicles. These are on the level of existence that we experience here – that is with our feet touching down in the physical world. This is the level that we can identify ourselves with being separate entities. The essence of the soul, or pure being, is at the centre of this arrangement.

I suggest we keep in mind that, however many vehicles we identify as belonging to the soul, the soul is not the same as its vehicles, and this is the point. Indeed, one can argue it is trapped in these vehicles until it can free itself from each in turn. It frees itself through learning and awakening – the opposite of how it got involved to begin with. Each vehicle, by the way, is also a creation of the soul that reflects where it is on its journey towards awakening.

The soul is the spark of eternal light. It does not exist as a thing, an object, in any spatial sense. All vehicles, we endeavour to identify, and including the soul itself, are transitory, an expression of Love – like a drop of water is to the ocean. An ocean is, after all, countless drops of water, if you want to see it that way. That might give us a clue too, that a soul can merge with other souls. That said, a soul in the bigger sense is immortal, is Love, but in outer expression is transitory.

It follows, as discussed in the previous chapter, that the separateness, and solidity, of things is ultimately an illusion. But given that, it best aids our understanding, application and progress in the physical world, where we currently exist, to view each of us as having a soul, that is separate and immortal, and as each being a separate entity. The Love's Story chapter will help to further clarify this concept.

The four vehicles are known as the "physical," the "etheric," the "astral" (or emotional) and the "mental." These run from lower to finer vibrations and frequencies, and are interwoven with each other, meshed together, though not in any way permanently connected. They are held in place by the soul.

I should mention that there is an obvious link here between the traditional perceived vehicles of the soul and the four elements of Western traditions: i.e., the physical with earth, the etheric with fire, the astral with water and the mental with air. A fifth element, "space," sometimes in this arrangement, becomes associated with the soul itself. Here's an overview of these vehicles:

The Physical vehicle: The physical vehicle is, of course, the vehicle we are so familiar with. It is the lowest vibration and frequency, and most dense of the four – made of matter, as we know and experience it. It is a beautiful and fine instrument for existence on the physical plane, with all the limitations and resistance it is likely to encounter or experience here. It literally is our Earth-suit for living on this planet, at the physical level of vibration.

The **Etheric vehicle**: The etheric vehicle, particularly at the level of the "etheric double," looks identical to the physical vehicle and is intermeshed with it as subtle luminous radiation. It operates at a much finer vibration than the physical vehicle. It includes the chakra centres and meridians conducting the life-force (prana, chi and kundalini energies) through the physical vehicle. The physical vehicle depends upon the etheric vehicle for its life, its health and vitality.

The etheric vehicle however must not be taken as a straight carbon-copy of the physical vehicle. Rather it is actually the other way around – it provides the template for the physical vehicle. It allows the life-force to pervade the physical vehicle for its duration. At death the etheric vehicle becomes detached from the physical vehicle. It is the vehicle we depart in, and with it is retained our appearance, our faculties and personality – all that we identify ourselves as.

The **Astral vehicle**: The astral, subtle, or emotional vehicle vibrates at a higher finer level than the etheric and physical vehicles. It is the giver of the life-force that sustains the etheric vehicle and more... Meher Baba describes it as, "*The vehicle of desires and vital forces.*"[4] This vehicle is immersed in what is known as the astral plane – the plane of phenomena where what is felt and desired appears immediately, or rapidly. Imagine it as like an ocean of subtle moods, light and sound, responding to every thought and desire. It is in motion with swirls, eddies and tides as feelings, desires, emerge and fall away again.

Here is a reason as to why we all need to practice meditation. Stilling these waters, particularly negative moods of anger, fear, worry and anxiety requires a watchful and discriminating approach to life, so that one is ever seeking to work with higher desires while inhibiting, and sublimating lower desires. This is the route by which the soul can be free of the physical vehicle, and eventually, the etheric vehicle.

The **Mental vehicle**: The mental vehicle operates within the mental plane.[5] It is the container of the mind and all faculties of thinking. It is made up of thoughts, just as the emotional body consists of desires and emotions and the physical body is made up of matter. In esoteric understanding, thoughts, once produced, have their own existence as thought-forms.

How we respond, think and produce ideas, through our mind, is inextricably linked to how developed we are in the other vehicles that make up the lenses of our soul. Our outer world experience and interactions in particular, help serve or hamper our spiritual growth and objectives. One's will is all important in harnessing and directing the mind towards higher thought and spiritually successful outcomes – outcomes that are expressed upon the mental plane and, in turn, filtered down through the lower vibration/frequency vehicles. The higher mind, the will, has the ability to create thought-forms that can bypass lower levels of desire and be expressed in love – thus generating transformation and refinement in the etheric and physical vehicles. We have to be careful not to let the mind wander though, but rather to focus it towards right intention – again, meditation can help with this.

The Light vehicle, or soul: Right at the heart of these vehicles lies the soul, in what is sometimes called the light or causal vehicle. From our vantage point (as there are, esoterically, more levels), this is the root of creation, of existence and the innermost essence of the soul. If one is to describe this as a vehicle at all (it is the fifth vehicle on that basis), it is otherwise the divine spark contained, and around which all other vehicles of the soul are built in the process of involution, and undone in the process of evolution. It can be called "pure being" and the interface between the outer physical expression and inner unity in cosmic Love.

In truth, as already suggested, the soul is only separated from Love by its belief in the illusion of being separate, by its level of awakening, in other words. It is, as mentioned above, a drop in the ocean of Love.

All beings have a soul

In my *Life and Death* book I asked if an ant could have a soul. The question is probably even more appropriate here, so I want to reiterate, in part, some of the comments I made there.

Would an ant have a soul, or is it just a little being that is filling a niche in the greater ecosystem, and completely the product of Nature and evolution? The simplest explanation would be to say the ant is the product of Nature and evolution and therefore does not have a soul. But to me it has a soul. Indeed, in my belief system it most definitely has a soul.

The ant, though, operates more on what I would take to be a closed collective instinctual intelligence, rather than by individual choices and decisions. Probably, like most insects, ants go to work with clear inbuilt directives – let's say an internal blueprint of what they have to do. This is not to say however that ants do not have the capacity to learn, as clearly they do (and need to), and have been observed, for example, showing each other the way to food sources. It's a case of the one who knows the way may have another follow in tandem, and they keep in touch with each other, so that the learner is keeping up.

The ants are hard-wired to do a range of specific jobs: the females are occupied in finding food for the nest, helping the colony to reproduce and be successful, but also to keep the nest shipshape too – and that means doing the housework. Males do very little, it appears, except to reproduce and die afterwards – and then get recycled, soul-wise, as I see it. The remarkable thing is that, in an ant colony no one is actually in charge (similar to bees) and yet they all work as one unit. The so-called "queen" is really a regular ant that has taken on the role of manufacturing ants for the colony, some task, but is not in charge of it. They instead know instinctively what they are doing, what role they are playing.[6] So I would say that while each ant is indeed a soul with a body, it is also part of a collective group intelligence that behaves like a group soul, no matter how closed and instinctual we choose to view it.

A spectrum of change

In my view, all beings, insects, plants, animals and humans are an expression of soul. It is stretching belief a bit, it would seem, but I would go so far to say that soul is present in all matter, arguably even more noticeably at the atomic and sub-atomic levels. Everything is undergoing change. Nothing remains completely still. Consider a spectrum, between the phlegmatic lower, slow to change levels, that we call the visible (that in other words is matter, or energy, and unconscious), to the higher and finer, fast vibrating levels that, to us, is invisible. Towards the middle of this arrangement we have what we would normally identify as life. Towards the higher levels we have what we could call the awakened mind or consciousness. The soul meanwhile can stand inside or outside of this arrangement – it runs through it. It could be likened to an agency made of Love with a directive towards Love.

Comment on Reincarnation and the Soul

My essential argument is that all life does have a purpose, and that purpose is tied in with spiritual development, with awakening, with consciousness – which naturally ties in with organisation, learning and development. Our planet is often described as a school for souls. Here we learn to negotiate our way from a much lower vibration and frequency, than is our true nature, towards the higher levels. This is learning through experience within physical bodies or vehicles, and with other souls, within linear time and working with the resistance such a place, as the Earth, offers.

In order for awakening to occur, in these circumstances, all souls must needs incarnate and have opportunity to reincarnate. Souls cycle and recycle therefore, and, as I'll discuss further in Part Two, all souls, at whatever level, head towards consciousness along a path of least resistance, in their lifeform expressions, until they achieve the form that gives them most opportunity to become conscious/awake – in their given planetary circumstances.

And these planetary circumstances need not be limited to our solar system, as we are discovering, and will be discussed in the next chapter.

Notes & references

[1] Note, I have written about the soul more extensively in "The Soul Question" chapter of Life and Death, the companion book to this one. This took into consideration the perspective of Christianity, Abrahamic religions generally, eastern religions that included Hindu and Buddhism, and esoteric perspectives. I am not intending to reiterate that discussion here. However, in context with where I am taking this work, it has been necessary to recap on a description for the soul.

[2] Drawn from The Different Bodies of Man
http://www.plotinus.com/subtle_bodies_copy.htm [Accessed 31/01/2018].

[3] The range of divisions can be into seven, nine or even ten vehicles or levels. In Theosophy there are seven levels taken together. These are: The Physical, the Etheric, the Astral, the Mental, the Buddha, the Atma, and the Monadic spark.

[4] Baba, M. (1967) Discourses 2 San Francisco: Sufism reoriented P145. ISBN 978-1880619094

[5] This is split into lower and higher levels. The lower linked to personality, logic, the higher to the intuitive, the abstract and spiritual – the creative, innovative, imagination.

[6] For more information on ants check out: http://en.wikipedia.org/wiki/Ant [Accessed 31/01/2018].

Chapter 5

Are we alone in the Universe?

I s it just us, all this teeming life on our planet Earth, or could this, even this Gaia, be happening out there too?

Whether you can go with Lovelock's original Gaia hypothesis or not, I'm sure you will agree, we live on a remarkable and beautiful planet. It is an important part of our solar system. At least we all think so – well I hope we do. We used to think that way...

This place is big

If one includes the Oort Cloud[1] our solar system is thought to be about two to three light years in diameter. If you don't know what that means, well, in simple terms, if we were travelling at the speed of light, 186.000 miles (299,792 kilometres) per second, it would take us roughly three years to cross it, from one side to the other. That is a pretty large distance.

As you'll know, our solar system is a part of the Milky Way galaxy. It orbits out in the suburbs of the galaxy. The Milky

Way is vast. It is estimated to be a hundred thousand light years in diameter. Without stating the obvious; travelling at the speed of light across this vastness would take us a very long time. It would be no stroll in the park, and our soul would definitely need to have a series of long physical lives if we wanted to complete the journey. But then this huge area is comparatively only a tenth of the size of the M87 galaxy. Here we are moving off the scale and talking of a star mass that is believed to be a whopping nine hundred and eighty thousand light years across.

No doubt about it, the universe is a very big place. When we try to contemplate the size of all of this star-filled space we are inevitably confronted with big star-sized questions. Or if not then we really ought to be. It is part of the adventure we are on.

Some time back I wrote a post (on SomeInspiration.com) with the same title as I'm using here, *Are We Alone in The Universe?* This is one of those star-sized questions. A question that probably a lot of us, inquisitive about life and its bigger concerns, might frame, and possibly we have done so going back many hundreds even thousands of years. In context it was of interest to me to discover that Winston Churchill posed the very same question, back in 1939. He wrote an eleven page document in answer to it too. This document was only rediscovered in 2016 (in the US National Churchill Museum archives). It was then picked up by astrophysicist, Mario Livio, who wrote about its contents in the *Nature* journal.[2] His article makes for an interesting read.

Churchill sought to answer this question from essentially a scientific and biological perspective – arguing the case for the probability of our not being uniquely alone in the universe based upon its size, and life here – its ability to "breed and multiply." I followed a similar path in my article but there, as here, I was also seeking to consider the question in context with a life imperative, intelligent life at that, and the spiritual implications that this could conjure for us all.

Another way of phrasing this question, "Are we alone..." might be to ask if this is really a unique, one-off situation we all find ourselves in here, or is it something that is likely to be more commonly replicated out there than we yet imagine? One answer to the question conjures a favourite quote of mine; by the astronomer, Carl Sagan:

> The universe is a pretty big place. If it's just us, it seems like an awful waste of space.

I think that says it all for me, and touches on our very human sense of humour; and sense of irony too, no less. Well, the evidence is mounting that space is not being wasted. We are unique, but not in our planetary circumstances – other planets exist that are similar. Neither then are we likely to be alone. And besides the evidence in support of this, we surely also know, in our belly, that this must be so.

Paradigm shift

But we are only just beginning to get to grips with this concept of life out there, around us. Consider that what we mostly take for granted now, our understanding of the world as a planet going around the Sun, is actually not all that old a concept. You probably know; it was once believed that we lived on a world held up by four elephants that in turn were supported on the back of a turtle. The turtle swam in a cosmic sea, and, by all accounts did pretty well considering the weight it was carrying.

We've come a long way since then, but it has taken some enlightened souls and frankly some serious arm twisting to get us here. It's not all that long ago that a great number of us believed that what we lived upon was fixed in the centre of the universe; possibly even flat (imagine the excitement in looking for the edge, and the fear of falling off it), where everything else, that is the Sun, Moon, planets and all those billions of stars, moved around us.

Much earlier, the Greek astronomer, Aristarchus (of Samos, about 310-230 BC), tried to convince us otherwise, that the Earth was round, a planet no less, and moving. He argued it was travelling around the Sun. We know that Aristarchus' theory went down like a lead balloon at the time. He couldn't compete with conventional wisdom, established by Aristotle, who never doubted the stars and planets circled the Earth. It wasn't until much later, the 16th century that Nicolaus Copernicus (1473-1543) dared to propose essentially the same idea as Aristarchus. Here we were again – some seventeen hundred years later – with someone arguing that the Earth moves around the Sun. Copernicus arrived at his conclusion in attempts to simplify all the weird and wonderful gymnastics some of the heavenly bodies had to make for the Earth to remain static in the centre of things.

For mainly political reasons (i.e., the dictates of the Church), his theory was also shot down. It might have remained that way, buried with Copernicus, had it not been for one or two influential supporters of his theory. These helped to keep his *heliocentric* model (Fig. 5.1) alive. He had the likes of the astrologer, John Dee (1527-1609), and the Dominican friar, Giordano Bruno (1548-1600), support him. By the way it is said, of Bruno, that he also had the temerity to suggest that the stars were distant suns, and further, could also have planets going around them, with life on them. What a leap in the dark for that time…

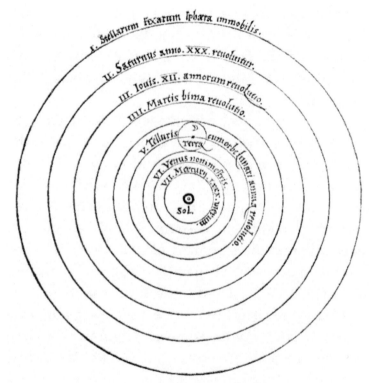

Fig. 5.1 Copernicus Heliocentric Model
Courtesy Wikimedia

The astronomer, Johannes Kepler (1571-1630), also supported Copernicus' theory, as did possibly the greatest of all well-known radicals, Galileo Galilei (1564-1642). If you remember, Galileo discovered the moons of Jupiter, and that they were orbiting Jupiter – not the Earth – so something was up. How could the Earth be at the centre of everything if this was going on? Well we know what happened to this heretic when the "Inquisition" caught up with him. The astronomer, Tycho Brahe (1546-1601), considered Copernicus had got the maths right but couldn't see how the Earth, *"that hulking, lazy body, unfit for motion, a motion as quick as that of the aethereal torches, and a triple motion at that,"*[3] could ever be for real. He suggested a compromise, a model wherein the planets went around the Sun, but then that the Sun, and the whole caboodle of stars, went around the stationery Earth. This was a clever theory and surely kept the Church happy, but it wouldn't last – it couldn't last.

Mad that they were, in their beliefs and observations, these souls took sides with Copernicus. Eventually, over another century or more, the truth did finally begin to get out; Copernicus' heliocentric theory did hold and was accepted. The Earth was deemed to be moving around the Sun.

But you see what I mean about the worldview we hold with now, not being all that long in the making, and even less so in acceptance. It is really only a few hundred years of acceptance. We have gone through a paradigm shift and today we know better. It's obvious isn't it? Well it is obvious to us now. The Earth is moving and indeed we know the whole lot more about the universe – especially that it is also moving. It is all moving and changing – nothing is as "fixed" as it was once believed.

> We are just an advanced breed of monkeys on a minor planet of a very average star. But we can understand the Universe. That makes us something very special.
>
> Stephen Hawking[4]

I would add; it occurs to me that the knowledge we have gained, since Copernicus, has also come at something of a price. When we believed the Earth was at the centre of everything, it was a pretty important place to be living on – even if based upon flawed assumptions. Now we perhaps don't have the same awe and reverence for our planet as we once had. It doesn't quite have that status now. Indeed, with what we now know, some of us have gone the other way, having become a bit blasé about the Earth's place in the scheme of things. Some of us like to remind the rest of us that we live on a pretty ordinary planet, the "third rock from the Sun" that is orbiting an ordinary star. Of course, if it had been the second (Venus) or fourth (Mars) rock from the Sun, we might have found ourselves in less palatable circumstances than what we are actually in. It could be worth reminding ourselves of how blessed we are. And besides, we are not quite ready to leave these shores, to move home – not just yet anyhow.

But to continue... Now, as we look out, from our spaceship Earth, with its Moon satellite, we are learning that this is a much bigger story to unfold than we might have thought. It is not just rocks, stars, dust clouds, black holes, dark matter and galaxies flying hither and thither, there is a lot more going on. It beggars the answer to the question (as per the chapter title), being a big "No." No, we are not alone in the Universe. With all things considered, it is really the only answer we can arrive at.

Where might we be in a hundred years?

Let's just remind ourselves where we are at now, in our own development. Besides accepting that we are living on a world moving through space, it is also fair to say that we have become technically quite advanced, in a relatively short period of time. Given this pace of change where do you think we might be in, say, a hundred years' time?

Let me make some suggestions: we could be more regularly visiting outer space. Indeed recreational space trips around the Moon are already planned, and within our grasp. And perhaps even one or two of the planets within our solar

---

system, perhaps Mars, could be visited by humans. In a thousand years' time could we not be travelling outside of our solar system? Or, if we, ourselves, are not doing it, then sending androids to other star systems? We might by then be occupying Mars or other planets within our system – and be meeting and greeting other lifeforms that we have gone looking for.

An even bigger paradigm shift

If we have cracked the space/time problem of travel, could we be moving from one location to another, light years away, more or less in an instant – well, okay, in a few earth minutes? The possibility is being taken seriously – using "wormholes."

Back in 2012 a NASA article, *Hidden Portals in Earth's Magnetic Field*, indicated that wormholes, portals, stargates, whatever you wish to call them, do exist. Okay, sort of, as this concept needs further exploration. The article draws on the research by physicist, Jack Scudder. In it, he comments that, *"We call them X-points or electron diffusion regions. They're places where the magnetic field of Earth connects to the magnetic field of the Sun, creating an uninterrupted path leading from our own planet to the sun's atmosphere 93 million miles away."* The article goes on to say:

> Observations by NASA's THEMIS spacecraft and Europe's Cluster probes suggest that these magnetic portals open and close dozens of times each day. They're typically located a few tens of thousands of kilometres from Earth where the geomagnetic field meets the onrushing solar wind. Most portals are small and short-lived; others are yawning, vast, and sustained. [5]

In my understanding, from an esoteric viewpoint, these portals could tie into the astral plane, providing what we'd call "astral travel." It has long been believed (some would argue, known for sure), that when outside of our physical bodies, on the *Other Side* as it were, in the astral plane, we can move around freely, instantaneously, by use of thought. So, the possibility arises that if we can find a way to transfer our physical selves, and our space vehicles, into this plane, via a portal or stargate, we can simply manifest at the target location. It's a thought that may become a reality sooner than we know it.

Could life be already visiting us?

If we, a couple of hundred years down the line from relying on horses as our main means of getting around, are already considering the possibility of space travel, via portals, would it be reasonable to assume that life out there has had similar thought, similar with regards to visiting other places from where they are?

We need to consider that we are not in the oldest star system going. There are other systems that are older, a lot older. And some of these older systems also contain planets in the "habitable zone" – the right location to support life (as we know it), as is the Earth. If they contain life, and if it has developed along anything like the pattern emerging for us, well…

My view is that, as life is in abundance all around us here, and, as we know from studies of the likes of extremophiles,[6] life is also adaptable to harsh conditions, and tenacious in its intention, it makes sense that, wherever possible, life is going to be teeming out there too. It will be just as intelligent, urgent and determined in getting established, surviving and reproducing as it is here. It's not rocket science, it's self-evident and around us.

The Milky Way Extraterrestrials

Back in 1961, the astrophysicist, Dr Frank Drake, and his SETI[7] team, set about estimating the possibility of extraterrestrial civilisations within the Milky Way galaxy. They came up with a guesstimate figure, a probability of ten thousand civilisations.

From what we know now, that potential could be much higher. If you have kept abreast of the news, you will be aware that, in 2015 NASA announced its 1000th confirmed "exoplanet" (a planet that orbits a star, stellar remnant, or a brown dwarf, other than our Sun), that had been discovered by the Kepler Space Telescope. That said it is common knowledge, with NASA's space telescopes, that there are indeed a huge number of exoplanets out there that are within the habitable zone – that could therefore contain life:

> In November 2013, astronomers reported, based on Kepler space mission data, that there could be as many as **40 billion Earth-sized planets** orbiting in the habitable zones of Sun-like stars and red dwarfs in the Milky Way. Eleven billion of which may be orbiting Sun-like stars.[8]

So, going by this guesstimate there could be possibly forty billion Earth-sized planets; and that number is estimated in just the one galaxy. Keep in mind that the visible universe contains at least a hundred billion galaxies.

On our doorstep

And could any of these planets be on our doorstep? Well, depending on how you view things, yes they could. Consider the TRAPPIST-1 system, discovered in 2013.[9] It has seven Earth-sized exoplanets orbiting a small dwarf star – with three of the planets in the habitable zone. It is nearly forty light years away from us, in the direction of the constellation of Aquarius...

Okay, so this is a huge distance in our current understanding of how one gets from A to B, but possibly not far by means we have yet to discover, or which have been discovered already by others living out there. And so, then I repeat, what are the chances that there will be life out there that is far more advanced than ourselves; as well as, fairly, the chances of the opposite, life that we can anticipate is far more primitive, if not in technology but in kind? I'd suggest that the likelihood is great on both counts.

And what about UFOs?

No doubt about it. In my humble opinion it is naïve to believe we are alone in the Universe.

Having gone this far I just couldn't leave this chapter without asking an obvious question. Against the backdrop of what is being covered, **do you believe in UFOs?**

Obviously I have no idea where you stand on this matter. I suspect that if you are sceptical you may have a bit of a negative reaction to this question. I suspect if I had asked you, "Do you believe in unidentified flying objects?" then you might more likely say you do believe there can be objects flying around in the skies that have not been fully recognised or fully identified, as yet. I reckon that most of us would say "yes" to the question on the basis that an acceptable or rational explanation can or will eventually be found to explain what is currently "unidentified" and "flying."

But if asking if you believe in UFOs; well the acronym here makes all the difference don't you think? It's potent. I'm sure you'll agree that when posed this way, the question takes on a different connotation altogether. Its link with extraterrestrial life (through extraterrestrial visitations - ETVs) becomes as good as given. And this tends to divide us into those who believe in aliens and those of us who don't. It can raise hackles too, as not everyone is prepared to even reason with the existence of aliens, or entertain the notion of their being, in any way, real.

Right now I'm thinking of an author, whom I know quite well, who adamantly doesn't believe in UFOs. An irony to this is that he writes sci-fi stories, and has even written a very good sci-fi story involving flying saucers and aliens. Outside of his writing though, he considers his feet are firmly on the ground where this matter is concerned. For him, there will always be an explanation for UFOs; that doesn't involve little green people from Mars, or anywhere else.

He's not alone, of course. Probably a great many of us believe that UFO sightings can be explained by planets, stars, cloud formations, weather balloons, drones, insects, sky lanterns, swamp gas, plasma, meteors, fireballs and the like. And this is not to overlook people also faking sightings, either by filming physical objects to look like UFOs, or by using computer generated imagery (CGI) to present very real-looking evidence. Either way, convincing but totally fake accounts can be spread around – particularly through social media. That said there is a good percentage of the public who do now believe in the existence of UFOs.

Regardless of public interest, either way, the establishment viewpoint tends[10] towards denial of their existence – calling on the kinds of explanations as listed above. But that still leaves a percentage of "unexplained" sightings. That being the case, even if there is a perfectly rational explanation for the unexplained – which is that extraterrestrials do indeed exist and are visiting our planet – it is easier, and probably safer, for us to follow the established response, rather than put ourselves in the firing line for being labelled as "gullible," and entertaining something, apparently, having no scientific basis, or that is unreasonable.

> Call me when you have a dinner invite from an alien. The evidence is so paltry for aliens to visit Earth, I have no further interest.
>
> Neil deGrasse Tyson, astrophysicist[11]

And, I could add that even a government body admitting an interest in UFOs doesn't mean an endorsement of them, in the way that we might imagine. You may have read, in the press, at the end of 2017, the US Pentagon admitted that it did run an "X-Files" department to investigate UFOs.[12] This department reportedly came to an end in 2012. I wouldn't count on that closure entirely though, and let me explain why: While it is tempting to suggest the existence of that department was an endorsement of a belief in UFOs and aliens (up the 2012 at least), we need to bear in mind that the Pentagon, and all governments around the world, especially with military capability, are interested in what enters their airspace. Unidentified flying objects, regardless of their origin, that poise any kind of possible threat, will always be of interest.

A lot of what is going on in the skies might well be explained by things that transpire to be quite familiar to us. However, even allowing for the familiar and the fake, the photographic/video evidence is becoming overwhelming that intelligent life, not necessarily of this world, well not the physical world at least, is behind some of what we are seeing. As the astronaut, Edgar Mitchell, once said on radio, and to camera, "*I happen to have been privileged enough to be in on the fact that we've been visited on this planet, and the UFO phenomenon is real.*[13] And he is not the only astronaut to comment on the reality of the phenomenon – Gordon Cooper[14] is another.

UFO sightings across the world, and especially in the US, are increasing almost year on year. The National UFO Reporting Centre (www.nuforc.org) figures show that between 2002 and 2009 there were over thirty six thousand sightings reported. Between 2010 and 2017 there have been over forty seven thousand sightings of UFOs reported. And these figures, of course, make no allowance for sightings that never got reported…which could be equally high – and I include myself in that category, as I'll explain below.

Three conditions

There are, in my view, three conditions we have to consider, and are obliged to accept, in order for us to take seriously visitations of extraterrestrial life to Earth. These have already been flagged up above but let me list them here more clearly:

1. The first is that we are not alone in the universe. There is life out there. This is becoming so obviously the case from the discoveries we are making. But it is also a logical and simple deduction if you consider how successful and diverse life is on this planet – which, like all other planets, is also in space (I think we tend to forget that). As I see it, life is urgent. It will occupy any environment, anywhere that can support it. It will always find ways of adapting. As it is teeming here, rest assured it will be teeming elsewhere in our universe.
2. Secondly, there is not just life but intelligent life out there. Some of the life will be older and more advanced than we – some of it is likewise going to be younger and less advanced than we. At this point we can say, we simply don't know for sure; although there are people who claim to have worked with extraterrestrials who, they say, are thousands of years in advance of ourselves.
3. Thirdly, if we accept the two conditions above, we can begin to accept that some of the intelligent life could be curious enough to be

visiting us – and may well have done so for thousands, possibly millions, of years. We make enough noise on this beautiful blue planet to let anyone in the neighbourhood know that we are here. It should not be a surprise therefore that it attracts visitors.

And fourthly, the big question that then follows from these conditions is, how are they getting here, or how did they get here if already here?

This is the logical stumbling block for a lot of us to believe in the existence of alien life (and UFOs) visiting us. And the answer to this question could well lie in the "portals," discussed above. If you stand back, and reason all of this, the only issue that can be described as a bit "woo-woo" or paranormal with these conditions, is the "how are they getting here" part. It is otherwise perfectly reasonable to consider that intelligent life out there exists, and could be, and most probably is, visiting us here. And not only may it be visiting; it may be living amongst us. There are huge tracts of land and seas where such things may be possible.

Two UFO stories

As if you couldn't guess, I happen to be one of those people, who believe in UFOs, of the extraterrestrial kind. I hasten to add that I didn't always believe in them. I'm also one of those who didn't report his observations at the time of seeing them.

Actually that latter comment isn't quite correct, I did query one with RAF Wittering (UK) back around the turn of the 1980s. This was at a time when I lived near Peterborough and pilots from Wittering airbase were on training, in Harriers, and flying over the country park where I lived. They were making a turn further on and heading back, presumably to the base. It was getting past twilight as I walked through the park, with the jets, one by one, screaming over. In context, I also noticed a bright light hovering above some trees on the edge of the park. I'd never seen a light at that spot before. To begin with, I thought that it must be some kind of observational point for the training, and probably a helicopter – although I couldn't hear any sound nor see any flashing lights – just a steady bright light. It was definitely hovering, gently moving up and down in relation to the trees, from my vantage point.

I decided that if I could still see it on my return home, through the park, I would give the base a call to check that they knew about it. Well, it was still there on my return, some twenty minutes later. So I did phone. The airbase had no knowledge of its presence or what it might be. Well, to be honest, I didn't think that they would tell me if they did know about it. I left it at that. At least I felt I had made an effort to report the sighting with them. I never saw any kind of light in that spot again.

Whether that was one, or not, I have certainly witnessed UFOs. Other members of my family have also seen them. And as indicated by the figures above, there's nothing special in that any more. I'm like thousands of people who have seen them. I have seen them on a number of occasions too – at least four apart from the one above. My most recent was in 2017, a steady light very high in the starlit sky, going over Evesham UK. That one was most probably a satellite.

Anyhow I'm going to share a couple of my UFO stories here that are more reliably of something other than aircraft or satellites we are used to. They are pretty ordinary against the encounters that some people have claimed, but they can be classified as unidentified flying objects, very probably of extraterrestrial origin. I consider they are germane to this chapter. Besides I'm going to enjoy telling them, so bear with me. These are both from the 1960s, so we're going back a bit.

Story one: A third light came hurtling...

Both these stories are going back to a time when I was involved in motorcycle racing. Like a number of the kids I grew up with, I was motorcycle-mad back then. The racing side of it though was more of a passion of my buddy, Eddie. It was Eddie's desire to race motorcycles, and later to own and drive a racing sidecar. This was ostensibly on grasstrack, but also on any other surface he could get onto, and he surely did when opportunity arose. It followed that, for a number of seasons, we raced together on racetracks around the UK – and, for the record, we did win some races too. Great memories...

As anyone involved in motor sport, particular at an amateur level, like ourselves, will know, there is a lot of tinkering to do between races – and especially between seasons. During the autumn, through to spring, you'd find Eddie and myself often working late into the evenings, repairing and preparing the bikes (we had two at one stage) for the new season. We'd be down at the shed, situated at the end of his garden. This was in Syston, in Leicestershire, UK.

The shed wasn't that large and it was normal for us to share working in or outside of it. As it happened, on this occasion in September 1966, I was doing my bit outside the shed. I was fixing something under torchlight. I looked up into the starlit night. It was a beautifully clear night. Our location, behind a row of terraced houses, meant that we didn't have a lot of ambient streetlight to contend with – only the shed lighting, and then really only when the door was kept open. So, outside it was getting pretty dark, except for the light from the torch I was holding. I don't know what made me look up at that particular stretch of sky when I did, but as I did I noticed a light, a steady light, like a star, that was moving slowly against the background stars. It caught my attention. I guess it must have been a few thousand feet up. Initially I thought it was a satellite – we were in the days of Telstar back then – and even if that had been what it was, it was a rare sight to see, and it gave me a buzz of excitement. As I watched this moving light, I called Eddie to come take a look.

No sooner had he poked his head out of the shed than another moving light had appeared and was joining the first. So now there were two moving lights, like stars, going in the same direction, but also appearing to be slowing to a stop. They then did indeed stop. We kept watching. Suddenly a third light came hurtling across the sky, seemingly out of nowhere but appearing from the same direction as the other two. It looked, for all intents and purposes, that the two had stopped to wait for the third to catch up. All three then went slowly off together across the sky in a southerly direction towards Leicester.

We looked on at the event as something really exciting. No planes could do what they were doing – and without any sound reaching the ground. We had no obvious explanation for it. It turns out too that we weren't the only people to notice these lights. A small piece appeared in the local Leicester Mercury newspaper shortly afterwards, about two people working on a street lamp having seen what we saw. There was also an account of it given in the LUFOIN Register (The Leicestershire UFO Research Society), which I only recently discovered when deciding to write about this. For the 3rd September 1966, the observation runs as follows:

3 blue circular aerial objects were observed that hovered for 20 minutes before drifting away: Three blue coloured circular objects were seen in the sky over Leicester, by two male witnesses while repairing a street lamp by the side of the road. [The] objects passed overhead and hovered for twenty minutes until [they] drifted away one after the other. [15]

The objects we saw didn't appear so much "blue" in colour, as I recall. It is possible that when we saw them they were high enough to pick up the remnants of the Sun's rays, so that what we saw could have been light reflecting off the craft. Anyhow, the rest of the description fits well with what we experienced. Bear in mind this was well before Chinese lanterns and drones.

I must say now that I wish we had also reported this to LUFOIN. Neither did I, or we, report the next story...

Story two: Heading towards the light

In a "Boys Own" sort of way, the second UFO event was much more exciting.

This event involved Eddie, his brother Malcolm, and myself. It took place around the same period as the first sighting – during the autumn of 1966. We were driving back from the other side of Grantham to Leicestershire one night in a van. All three of us were in the front seats. At the time, we used this road, the A607, to the other side of Grantham and back a lot as we had friends involved in motor sport over there – and I also had a girlfriend over there too. On this occasion Eddie was driving. It was late, probably around midnight, and we weren't much up for talking about anything. We were rather more focussed on just getting home, twenty-six or so miles away.

As you drive a short way out of Grantham, you travel up a gentle snaking hill rising onto, what one might describe, as a level of fields and farmland that runs for a few miles – nice countryside. As we did so this time, I noticed a bright light in the sky, in front of us, relatively low on the horizon, and seemingly in a fixed position. It was also a fairly cloudy night so I knew this wasn't a planet or star. I wondered if it could be the top light of the Waltham-on-the-Wolds radio mast (the mast was over a thousand feet high and, as you'd expect, it had four or five aircraft warning lights running up it). But Waltham was around 10 miles away and I'd never noticed any of its lights from this distance, or angle, before.

However as we continued up the road, and bearing to our right, the apparently stationary light appeared to move around to our left – well away from the direction of Waltham. I knew then that this was no light from a radio mast. I kept an eye on the light which now seemed to stay with us, low on the skyline.

Thus far in our journey I hadn't said anything to the brothers about it – nor had I broached anything else for that matter. As I said, we weren't up for much talking. When I did eventually break the silence it transpired they were also aware of it. They had picked up on it and were as interested as I was in what was going on. It soon developed into a topic for discussion, particularly as it appeared to be following us. Well forget the "appeared to be following us." It was, most certainly, following us, and did so for miles. It kept low in the sky and, seemingly even followed our every contour in the road.

We discussed what we should do about it. We were concerned to get home, but this was something else and couldn't be ignored. We came up with a plan. Here's the "Boys Own" bit. We decided that if this light stayed with us beyond Croxton Kerrial (about seven miles from Grantham), we would take full advantage of the big dip in the road, on the other side of the village, to give us momentum to go "full-chat" through the sharpish left hand bend that followed, probably about half a mile further on from Croxton. We would then, in theory, be driving towards this light. If circumstances permitted, we would even drive under the light and stop.

Well, the light continued to track us. We drove through Croxton and continued on our way in the direction of Waltham and Melton Mowbray. As we went through the valley dip in the road so again the light also seemed to take the same contour in its movement. We continued on through the left bend as fast as we could, and now we were, as we anticipated, heading towards the light. The light, meanwhile, had stopped and now remained stationary in the sky. It looked to be only a few hundred feet in the air, if that. All fired up, we stopped directly below it. We leaped out the van and stared up at it – excited and wondering what was going to happen next…

Well, disappointment; nothing much really. It remained stationary momentarily and then without a sound it moved on at an accelerating pace away from us.

One minute full of expectation, and the next deflated that nothing had happened. We got back in the van and drove home. Either way, the event nourished our imagination, certainly mine, and I for one was glad I had witnessed it. Events like this encourage us to ask those bigger questions of life. The object looked like a simple bright light to us – a bit like the leading light in one of the chases from *Close Encounters of the Third Kind*, but this was way before the movie.

As I said earlier, we didn't report this event at the time. To be honest it didn't occur to me to do so back then. I wasn't aware that anyone was compiling this information. The only record that I've found of any possible UFO along that same stretch of the A607 is again in LUFOIN:

A stationary bright light was observed by married couple … "Between mid-December 2012 and 28th Jan 2013 … my husband and I have travelled 3 or 4 times in the evenings on the A607 between Waltham on the Wolds and Melton Mowbray in both directions and every time we look in the direction of Scalford/Holwell we see a stationary bright light...[16]

I'm with Edgar Mitchell in saying there is "no doubt in my mind that the UFO phenomenon is real." And it has been real for years no matter how much we try to cover it up. Note I pick up again on extraterrestrial matters in the Epilogue, which also carries some links that you may find useful.

We live in uncertain yet exciting times... don't you think?

Notes & references

[1] Oort Cloud, an extended shell of icy objects believed to be on the very edges of our system, spherical in shape, and where comets are believed to originate.

[2] Nature. February 2017. Vol 542. Pgs 289-291

[3] Gingerich, O (1993) *The eye of heaven: Ptolemy, Copernicus, Kepler*. American Institute of Physics, 181. New York.

[4] Professor Stephen Hawking, TED talk, 2008.

[5] Scudder Dr T. (2012, June 29) Hidden Portals in Earth's Magnetic Field. NASA Science Beta. https://science.nasa.gov/science-news/science-at-nasa/2012/29jun_hiddenportals/ [Accessed 31/01/2018]

[6] Extremophiles: Such as the red flat bark beetle, surviving in temperatures down to minus one hundred and fifty degrees Celsius, or the desert ants of the Sahara, who daily do a "fire walk" in sixty degrees Celsius to obtain their food, or the Himalayan jumping spider who lives so high up the mountains, where no other life exists, that it relies on the mountain winds to blow frozen insects onto its lair. Source: http://www.bbc.co.uk/nature/21923937 [Accessed 31/01/2018]

[7] SETI Search for Extra-Terrestrial Intelligence. Visit https://www.seti.org/

[8] https://en.wikipedia.org/wiki/List_of_potentially_habitable_exoplanets [Accessed 31/01/2018]

[9] Visit http://www.trappist.one/ [Accessed 31/01/2018] to find out more on this system.

[10] I say "tends" here as some critics, for instance SecureTeam10, suggest we are being subjected to a drip-feed acceptance, that attitudes are changing.

[11] Ellefson, L (December 21, 2017) Neil deGrasse Tyson on UFOs: 'Call me when you have a dinner invite from an alien' CNN News http://edition.cnn.com/2017/12/20/us/neil-degrasse-tyson-ufos-new-day-cnntv/index.html [Accessed 21/12/2017]

[12] Actually the Advanced Aviation Threat Identification Program. Example of reports: read Alex Hollings (2018 January 15) *SOFREP's X-Files: Four things to know about the Pentagon's secret UFO department*. SOFREP News. https://sofrep.com/98036/sofreps-x-files-four-things-know-pentagons-secret-ufo-department/ [Accessed 20/01/2018]

[13] On July 23, 2008, Edgar Mitchell was interviewed on Kerrang Radio by Nick Margerrison. Mitchell claimed the Roswell crash was real and that aliens have contacted humans several times, but that governments have hidden the truth for 60 years, stating: "I happen to have been privileged enough to be in on the fact that we've been visited on this planet, and the UFO phenomenon is real." In reply, a spokesman for NASA stated: "NASA does not track UFOs. NASA is not involved in any sort of cover-up about alien life on this planet or anywhere in the universe. Dr Mitchell is a great American, but we do not share his opinions on this issue." See Daily Mail http://www.dailymail.co.uk/sciencetech/article-1037471/Apollo-14-astronaut-claims-aliens-HAVE-contact--covered-60-years.html [Accessed 31/01/2018]

[14] See *An Astronaut's UFO Experience - Gordon Cooper*. Sirius Disclosure Exclusive (on YouTube) https://www.youtube.com/watch?v=wsEd_b1C8DY [Accessed 13/03/2018]

[15] Quote from The LUFOIN Register (of The Leicestershire UFO Research Society), Report Review 1954-1977. See 3 September 1966. http://lufoinregister.angelfire.com/B.htm [Accessed 31/01/2018]

[16] Quote from The LUFOIN Register (of The Leicestershire UFO Research Society), *51/13/07*. 28 January 2013 - 20:00 - Scalford, Leicestershire, UK - A stationary bright light was observed by married couple. *Witness (A) - Wife - stated:* "Between mid December 2012 and 28th Jan 2013 (last night) my husband and I have travelled 3 or 4 times in the evenings on the A607 between Waltham on the Wolds and Melton Mowbray in both directions and every time we look in the direction of Scalford/Holwell we see a stationary bright light that at first we thought was Jupiter or Venus, but as we watch it, it changes from a bright yellow light as we drive along, to flashing green/red like an aeroplane then changes to yellow (not flashing) again. We can watch it for a good few minutes and then it seems to disappear. Waltham transmitter is in the other direction while we observe this. If it is a hovering helicopter why is it in the same place every time we travel this road whatever the time of night? It can't be a star as it changes colour but not position in the sky, and not a plane as it does not move. We think it may be another transmitter but a long way off and so can't see a tower below it?" http://lufoinregister.angelfire.com/G2.htm [Accessed 31/01/2018]

Part Two

This part is where the nuts and bolts, and discussion on Love's Agenda hypothesis, of *why we are here*, takes place – and also includes exploring *what we can do about it.*

God sleeps in the rocks
Dreams in the plants
Stirs in the animals
Awakens in Man

Ibn Arabi, 12th century Sufi teacher

Chapter 6

Awakening

G iven that there are a huge number of planets that could contain life in the universe; and given that the universe is very probably teaming with life, as our planet is likewise, a question arises: Is there a bigger picture in place that we need to grasp? Could there be a spiritual purpose to life that involves not just us but all lifeforms? Could there indeed be a grand plan in operation behind life after all?

In the following pages, and chapters, I will be exploring, what I believe to be a plausible "grand plan" agenda operating behind life, with a view to arriving at a description for it – indeed to propose a possible agenda operating not just here, on the Earth, but elsewhere in the cosmos.

> Recognise that the very molecules that make up your body, the atoms that construct the molecules, are traceable to the crucibles that were once the centres of high mass stars, that exploded their incredibly rich guts into the galaxy, enriching pristine gas clouds with the chemistry of life. So that we are all connected to each other biologically, to the Earth chemically, and to the rest of the Universe,

atomically. That's kind of cool. That makes me smile… we're part of the Universe. We're in the Universe and the Universe is in us.

Neil deGrasse Tyson, astrophysicist[1]

I enjoy these inspiring words of Tyson. It makes me smile too. The idea of the Universe in us and us in it, that's a "wow." On a profound molecular level that connects us to everything that is around us. It sort of requires a deep breath to take it in.

There is of course the danger that we'll respond to such an observation with something like, "That's interesting!" and then leave it all to one side as being either too abstract, or really too big for us to feel or comprehend. And we move on with what we have planned for today, or for tomorrow – what we call, "getting on with our lives." I think you would agree that it is very easy for us to become detached in this way, not just from the bigger picture but from much of what is going on in the world around us. But I tell you – we must endeavour to go the next step… It really is that important, for we are in it, this life-stuff, and it is within us; and it really should excite and inspire us to get to grips with it. We must get to understand this and indeed ask if, in context with being a part of all of this wonderment, there is also a purpose to our being here. It is a measure of our wakefulness.

It surprises me how disinterested we are today about things like physics, space, the universe and philosophy of our existence, our purpose, our final destination. It's a crazy world out there. Be curious.

Stephen Hawking

A purpose to life

To return to my earlier comments, relating to Gaia and evolution. I don't think it hurts to repeat that from the conventional biological perspective, life on Earth is believed to have been all hit and miss at the beginning. Once it got underway it brought us, eventually, to the world of flora and fauna we see around us. From this position it is easier for us to conclude that there is no purpose to life, or that the purpose is to respond to, make the best use of, the immediate circumstances we are in. We therefore compete or cooperate with each other, and with the environment, and adapt as best we can. We, like every other lifeform, are thus merely a part of all the multi-various forms of life on the planet.

But, as I've pointed out, the way life is organised beggars that we look deeper than this and consider the possibility of motivation and purpose beyond this comparatively simple but, excuse the pun, lifeless explanation for evolution held by conventional wisdom and science. It doesn't take much to notice that there is intelligence and order operating everywhere; and this intelligence and order leads one easily onwards to consider that this is orchestrated for something other than just existing from day to day, week to week and year to year. I suggest that however life may have started out (for my money with "intention" from the get-go) it has gone well beyond the "surviving to exist" point for a lot of it. It is too co-ordinated, too balanced, too wilful, too beautiful of form, too intent on survival that there has to be something more to this.

Well, you know what I'm going to say; there is "something more" to existence in my view. It's a "something," I believe, with a little more than casual observation, to be self-evident and impossible to ignore.

Hierarchy of needs

To aid my discussing this further, I'm going to draw on the elegant theory proposed by the American psychologist, Abraham Maslow. This is Maslow's "hierarchy of needs" (Fig.

Love's Story Of Why We Are Here

6.1) which he first presented back in 1943,[2] and has been a popular model ever since. The model is designed to help our understanding of how we humans might best achieve a healthy psyche and make the best of our lives. The five levels draw on the notion that self-actualisation, or fulfilling our full potential, at the top of the model, rests on having in place, and successfully mastering, the requirements of the four supporting lower levels. These levels, each in turn, provide the foundation to our fulfilment.

Fig. 6.1 Maslow's Hierarchy of Needs

The theory goes that:

Level One: Given our most basic (physiological) needs for food, water, shelter and clothing are met.

Level Two: Combined with our needs for stability, being in a safe and secure place, are also met.

Level Three: We can more easily develop a sense of belonging. We can feel more "at home" within our surroundings, with our family, friends, with society, and especially within ourself. We can feel loved.

Level Four: Which, if met, in turn gives rise to feeling good about "who we are," being at ease with our identity. This has the potential to lift our self-esteem and has the power to better help us find, and establish, our place in the world.

Who of us wouldn't embrace the promise of each of these steps, whether in actuality we received none of it, some of it, or all of it.

These four steps act like pillars to provide the very foundation for realising, or achieving our full potential. We can become a fully loving, mature and (in the words of another well-known pioneer of humanistic psychology, Carl Rogers) a "fully-functioning person," by applying them, or having them applied in our lives. And it is not to be limited to individual endeavours either; we can look at the bigger picture of societies and cultures that do or don't (or can't) adopt and embed these principles into their structures. They succeed, or are more likely to fail, if they don't.

Both Maslow and Rogers held that all living things naturally strive towards self-actualisation, self-fulfilment. This is what Rogers called the "actualising tendency." It is a beautiful idea in itself. Let me give you a simple example of this tendency, from his book, *Carl Rogers On Personal Power*. Drawing on memories of his childhood home, he writes about what happened to a bin of potatoes left over winter in the basement – and several feet from a small window:

> The conditions were unfavourable but the potatoes would begin to sprout – pale white sprouts, so unlike the healthy green shoots they sent up when planted in the soil in spring – but these spindly sprouts would grow two or three feet in length as they reached for the distant light of the window.[3]

I suspect if you have ever stored potatoes in a garage, shed, basement or even a cupboard, you can probably relate to this story. I've certainly seen potatoes sprouting in similar circumstances. The point being made here though, is that clearly as the necessary conditions were not all in place, these potatoes were not going to fulfil their full potential. But yet, by the fact of being alive, they were taking a good shot at it. *Everything strives to actualise.*

Not necessarily all plain sailing

It has to be said, referring to Maslow's hierarchy of needs, there is no certainty that, having all the necessities for a well-balanced life in place, we will achieve our full potential. Using a sailing analogy, these four lower levels, let's say, provide the sailboat and support influencing the vision and compass of opportunity for a fulfilling life voyage. In that respect the template provides insight into optimum conditions for the best outcome; that is the sea-worthiness needed to arrive at our destination – whatever destination we may have chosen.

All levels being fulfilled, all set, and having the confidence to take the helm; one still has to set the sailboat on course. We have to do it on the spot, so to speak. We cannot, therefore, rule out the need for, *opportunity, co-ordination, application, willpower, self-belief, intention, motivation, purpose, perseverance* and, no doubt, a few other conditions also being necessary requirements, to the direction and the success of our endeavours. And, I think it fair to say, that given we have the necessary confidence, this resource we mostly draw on from within ourselves. Indeed, that said, the opposite conditions may apply; that regardless of our success or failure at acquiring the "necessities of life" to support our endeavours, we can, through our own "tour de force," still end up successful in what we aim for. We may start our journey on a simple raft (modest beginnings) and find, with the wind behind us (fortune in our favour – a helping hand maybe), we are able to make the best use of it on our voyage. We succeed in, say, next finding the boat we always wanted, or needed to find, in order to take the next leg of our journey to success.

Life and Consciousness model

The one begets two, the two begets three, and three begets the myriad creatures.   LaoZi

Given the above, let us now look at another version of a five level template (Fig. 6.2), that I have adapted in the style and direction of Maslow's model. This one is set up with the aim of displaying a much broader picture. I'm calling it the Life and Consciousness model. It is to illustrate a macro picture for life. I'm using it to illustrate a "proactive evolution" of life, from an unconscious state to self-consciousness or awakening state – in five broad steps.

This is intended as by way of giving an overview of the process of evolution, as I see it, from the starting pistol all those billions of years ago. This is not to overlook that what we might call primitive evolution is also happening now as ever – such as can be evidenced by the spontaneous growth of hydrocarbon molecules around the vents of the "Lost City" hydrothermal field in the Atlantic Ocean. This, very rudimentary life, is happening now and spontaneously. If you haven't heard of it, take a read of the Wikipedia version.[4]

I propose that there is an *invisible intelligence* (that I'm calling Love) that is the imperative driver behind life. This life and consciousness model is not intended to be scientific, by the way, rather philosophical and speculative. The model is, by its nature, simple but in being so it serves my purpose of getting across the concept that I'm aiming at – which is also actually simple. It is a starting point in my quest to suggest an answer to the bigger concern of, "Why we are here."

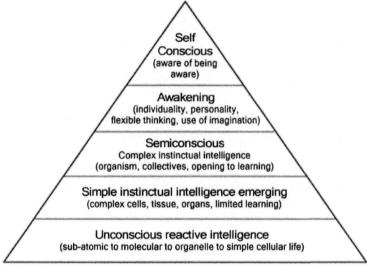

Fig. 6.2 Life and Consciousness

Let's take a look at this template, starting from the lowest level:

Level One: Unconscious state. Life is sparking at a very basic and reactive level. The invisible intelligence is in the early stages of its work, and any output it achieves is at a rudimentary unconscious state. Things are at their beginning. It starts from the sub-atomic/atomic, then to the molecular, then to the possible LUCA, but we know it develops into bacteria, then to the organelle, and to the eukaryote cell life. There is fundamental chemical interaction and symbiotic activity taking place.[5]

Life is firing at every opportunity – or miss-firing depending upon the circumstances or conditions in the environment. There is, at this stage, a concern with initiation and establishing the most basic needs for survival. I'm suggesting this corresponds with Maslow's, "Physiological needs" level.

Level Two: Instinctual state. At this next developmental step, life is in a position of getting organised. The intelligence is manifesting in growth and movement at the instinctual level. Complex cellular structures, involving tissue and organs, and new lifeforms are beginning to emerge in water and on land. Life is experimenting, going out in all directions. Trials and experimentation are taking place in response to already existing lifeforms and the environment. Communication and symbiotic relationship is at a fundamental level. Life is very much finding its feet and learning as it goes. The learning in this condition is limited to organisation, securing and consolidating its position. Here too, life is also beginning to encroach upon life.

I'm equating this state to Maslow's "Safety/security" level – a preoccupation with having the necessary conditions and protection to endure and thrive in the environment.

Level Three: Arousing semiconscious state. Here the intelligence expressed through life is at a semiconscious level. Life is speeding up and branching out. It's the emergence of more complex and larger organisms trending towards becoming more streamlined and sophisticated in form. There is greater communication (internally and externally) and symbiotic relationship taking place – invoking a more complex transfer of behavioural outcomes involving innovation and learning. There is a blossoming of collectives and diversity in plants and animals. Here there is growing scope given over to new learning and organisation – in turn requiring more time and opportunity for organisms to open up to new experience and to absorb it.

This level I see as equating to Maslow's "Sense of belonging" or territorial ownership.

Level Four: Awakening state. Here begins the emergence of individuality and personality. At this step, the intelligence behind life is becoming manifest, and gaining a degree of awakening through the lifeforms developing. It is also marking greater freedom to think and create, and to access imagination. Higher mammals, such as whales, elephants, and chimpanzees appear to operate in this state; as do bats and birds (such as the crow family for example). Any organism achieving this position is in an awakening transformative state, beginning to move out of, what I would call, a *state of innocence* (unconscious to semiconscious states), that is otherwise binding on other flora and fauna, towards a state of wakefulness and potential freedom of thought and individual action.

I equate this to Maslow's "Self-esteem" level. At this level though, without full awakening, individuality and personality will tend to remain limited at a cognisance level that is local and probably best described as "childlike."

Level Five: Self-conscious, awake state. This level is the pearl in the oyster that I see as equating to Maslow's "Self-actualisation" level. The previous steps all lead to the opportunity for the intelligence, operating through life, to become visible, fully awake, self-conscious, at the pinnacle in this scheme.

This is where life, through its lifeform/s, becomes aware of itself, aware of being aware, conscious of its own mortality, becomes conscious of its environment, and conscious of its place in its environment. It has access to express and experience imagination, thought, feeling, intuition and sensation. It embodies the concept of play and creativity. It not only objectively and subjectively interacts with its environment but also creates (pre-emptive) interaction and communication with its environment, and the bigger world around it. And it is now aware and responsive to the outcomes of its interactions.

This top level also entails a growing responsibility and maturity in extending consideration, compassion and positive regard to life, and all creatures – creatures that also originated from the same life source, are thereby are all related and seek the same awakened outcome. In this model, all lifeforms help provide four essential layers, or pillars, for the emergence of the awakened form.

The human form

It is fair to say that currently this fifth level, on the Earth, really only applies to human beings. Other creatures are not far behind but may never, in their current form, achieve self-consciousness with all the ramifications and dynamics that this state entails. Apart from the obvious, that it is not happening, it

---

is my view that a huge number of lifeforms are not designed to achieve a fuller level of consciousness, in the state they are in. Rather they serve as part of the experiment and structure supporting the awakening. But that is not where their story ends either...

Awakening is a difficult process. If we want the freedom it offers we are obliged to step out of our comfort zone of innocence and forgetfulness, and take the plunge, so to speak. As we can see around us, it brings the "us and them" relationship, the territorial imperative, we can have with our world, into much sharper focus – with potentially more selfish interest in, and design on, what others have, or don't have, that we want, as a result.

Once in this awakened state we are more aware of others effect upon us. We become aware of ourselves in the story. Our initial tendency is to look after *numero uno*. We can easily spend a good part of our awakening time following the old instinctual ways of doing things; and taking two steps back for every three forward. Only later in our awakening, when we have in place the pillars of Maslow's model, are we likely to begin to see that we are "all in this together;" and can extend the hand of love and friendship to one another – indeed to become the custodians of the life we share in.

I don't think we need to look far into our human endeavours to acknowledge that being awake has had (and still very much has) its risks. A being that has effectively been asleep and now becoming awake is potentially (in the early stages) a danger to itself and other forms of life around it. It's a risky venture and it probably takes time for the dust to settle, for any form that becomes self-conscious.

Because of the natural interdependence of plants and animals upon each other (in, for example their symbiotic support, the food chain and reproduction), you can imagine the likely outcome if all lifeforms achieved such wakefulness, and rapidly. With everyone looking to preserve their own interests, while probably seeking to take advantage over the interests of others, the whole ecological system could be in jeopardy of breaking down. Better to experiment with one or two forms, having good design and capacity to awaken, than to let everyone through that hoop. Consciousness is probably also only possible, or impossible, by the cellular wiring of a given form.

Although self-evident, it is reasonable to suggest that the most likely form for life to achieve consciousness is the bipedal, anthropoid form that we humans exist in. It is one that is most adaptable for developing tools and applying its learning to increasingly larger endeavours. In the business vernacular, it is designed to be able to "scale up" in other words.

In context, my guess (and I'm sure I'm not alone in anticipating this), is that this bipedal form will be endemic throughout the universe. It does not therefore surprise me that descriptions given by people, who claim to have seen intelligent life from elsewhere in our galaxy, describe the configuration of their form (head, body, two arms and two legs – each having a number of digits) as not that dissimilar to our own.

A problem with our current spiritual beliefs

By now then, if I haven't telegraphed it enough, you will have sussed that my raison d'être for life on this planet, is achieving "consciousness;" that is being awake, at its pinnacle. But there is more, much more...

From this juncture, I want to begin considering consciousness in context with spirituality. To do this, I also need to pause for a moment and consider our human situation as it is understood from a religious perspective. To pick up from where I left the soul (in Chapter 4), let me say that it doesn't matter whether you come from the Earth, Mars, Alpha Centauri, another part of the Milky Way, or Zardoz (remember that 1970s sci-fi movie?), you are a soul, or spirit, if you'd prefer. Whether you are human, an extraterrestrial, a plant, flea or a toad, you are a soul first and foremost. Indeed I would say if you are a rock you are also made of the stuff of soul, admittedly at a very low vibration and frequency, lying dormant, waiting to be recycled, awoken back into life. By the way, I love the Ibn Arabi quote that I began Part Two of the book with. Replace "God" with consciousness or soul and we're talking the same language.

All that said, our coming, going, and returning here on the Earth has long been believed, for thousands of years, particularly by those of us who hold with reincarnation or rebirth, as the way the situation works. What I mean is that we souls arrive here, taking up human form, and through gaining consciousness, we eventually lift ourselves out of the need to return here. We break free of the bounds of samsara, or desire, and that is it – we've reached Nirvana, or something similar. Within the Abrahamic religions, of course, it is even simpler, as we only have the one life, again in human form, before, at death, we are judged as to where we belong.

Clearly, to complicate matters, with the growth in population one has to also consider that souls, probably new souls, arrive here, year on year, and join in the birth-death-rebirth process – with the same intention to eventually break free of desire.

Lifeforms, other than human, meanwhile are assumed to play little or no part in this spiritual process. They are believed by some to be here simply to help provide the environment, and nourishment, needed for our development – which in part, of course, is true – there wouldn't be an ecosystem without them. There can be some degree of inclusivity of other lifeforms, if, for example, our religion is Eastern in origin and we believe our next incarnation could be a lifeform other than human – perhaps a tiger, a buffalo or a bird. I don't personally buy into this belief in the way it is presented, as I trust will become evident below. However that is somewhat beside the point I wish to make here.

What I am intending to demonstrate is that there is something of a conundrum with this birth, death, rebirth explanation as it stands. Given the comments above, a couple of queries arise that I see as needing additional thought and clarification. These boil down to:

1. Why should we assume that gaining our spiritual freedom from ties to the Earth, whether by a series of lives (Hindu/Buddhism) or by one life (Abrahamic religions) is all that life is really about for us?
2. And, in context, could there be a bigger, collective, story to this, a story that is being missed, that we are also a vital part of?

Let's look at the first query: In part, what I'm getting at here is that for many of us, who believe we are spiritual beings, also believe we arrive on the Earth and only ever take on human form. This may be true, but I would suggest this will only be true once we are ready; once we are able to vibrate at the level and frequency required to appear in human form. A great many of us will meanwhile exist in other lifeforms that make up the flora and fauna of the Earth. That said it may also be true for some of us to have spent much of our time elsewhere, in the universe, before arriving here and taking on human form. As already discussed, the Earth is not going to be the only place on which life exists, and some souls could experience, or be seeking to experience, lives in different environments, on different planets.

In context with this belief, it is often assumed that in essence we souls are, or were already, pure when we arrived here. If this is the case then, in coming to the Earth for the first time, we did (and obviously still do) run a very big risk. This is the risk of getting ourselves ensnared in the vibration of the physical world – the world of desire, getting ourselves caught in the cycle of samsara.

If you follow this logic; this means we have allowed ourselves to become trapped here, and are having to then seek out our freedom. Now if we also reason why we originally came here (and are still coming here) is *to experience, to learn,* one might fairly ask, was/is the risk actually worth it?

You see, our religions and spiritual systems are for the large part set up around helping us to lead a good life, in order to either enter heaven, or free ourselves from our trapped situation. So okay we may put it all down to experience and learning, to be good, or to work successfully with our karma. But that might be the same as saying we put ourselves into difficulty, into this mess, and are now having to learn our way out of it.

Perhaps needless to say, I now have some difficulty buying this perspective as it stands. This book endeavours to put forward a different model, or take, on this matter.

In my second query: Rounding out and freeing our soul, through experience, is surely an understood intention but are we not also a part of a greater whole of life? What I'm getting at here is: could there be more going on with this life story than simply our (human) soul salvation?

Well let me say I've not found a simple answer to this concern, either in the spiritual traditions I feel akin to, nor via people with esoteric knowledge, whose books I have read or who I have spoken to. Meanwhile I am fully signed up to the view that all of life, whether animate or inanimate, is a manifestation of Love.

To my mind, you see, it is *Love* that drives the system ultimately; that is the hidden intelligence in my model. If I add into this mix, a belief that all of life is undergoing experience that is *meaningful*, I have to consider there is a bigger agenda operating behind our being here on the Earth than just our own salvation. I'm going to suggest the answer lies in what I have already begun to describe above, in what I'm calling, "Love's Agenda."

Heading for wakefulness

Consider that the scenario I've thus far painted is not the whole picture. We are here not so much to deal with, and escape, the treacle, the resistance we individually experience being in physical life, but we are here on a mission. We are here on a challenging and dangerous, yet rewarding, mission. Consider that while some of us souls, who manifest here as human beings, have become fairly mobile – coming here and going elsewhere in the cosmos, or other possible dimensions – most of us have evolved here and have continued our spiritual journeys on this same planet. We have developed from simple to more complex lifeforms, following an evolutionary path.

The reason for this, I suggest, is that all souls, or in other words, all lifeforms, are required to find their niche (which depends upon their current vibrational state), then support the greater good, participate in the mix, and head for wakefulness.

That said, as is evident, the majority of lifeform routes to "wakefulness" are limited. On this planet we therefore mostly (not without resistance) seek to migrate up the evolutionary ladder towards the human level, the most promising route, to where we can become awakened, become conscious. At the human level our remit is to begin working with Love and compassion. To continue to be life and to help all life – those of us who are animals, plants, elements and rocks – to awake, to speed up the process, to indeed help the Earth speed up its desire to awake – as we also achieve being amongst those who have become conscious and mobile; and can come and go at will.

This is what makes sense to me, and I hope it will to you too, as I develop out Love's Agenda argument through the following pages – see what you make of it.

Notes & references

[1] Neil deGrasse Tyson, Astrophysicist, American Museum of Natural History.

[2] Maslow, A. (1943) Paper: *A Theory of Human Motivation*. Psychological Review, 50 (4)

[3] Rogers, C. (1978) *Carl Rogers on Personal Power*. Constable London.

[4] Lost City Hydrothermal Field. For overview see https://en.wikipedia.org/wiki/Lost_City_Hydrothermal_Field [Accessed 20/11/2017]

[5] Life functions through the specialized chemistry of carbon and water and is largely based upon four key families of chemicals: lipids (fatty cell walls), carbohydrates (sugars, cellulose), amino acids (protein metabolism), and nucleic acids (self-replicating DNA and RNA). Wikipedia https://en.wikipedia.org/wiki/Abiogenesis#The_deep_sea_vent_theory [Accessed 14/11/2017]

Chapter 7

Love's Agenda

"It's Life Jim, but not as we know it!"

C ould there be an agenda by, what I'm calling Love, that is the driver behind existence, that is the source where the buck stops for our manifestation?

Whether the Gaia Theory can be finally accepted or not, in the full sense that Lovelock originally envisaged, I embrace the idea of it, as it does make complete sense to me and supports my own speculations here. Gaia alive, having active intention, necessarily features in my understanding and the hypothesis I'm proposing.

Take a look at what I am proposing here. I have organised it into seven principles. What I am saying is, of course, speculative, but I would also suggest logical. It is similar to the "strong forms" of Gaia (discussed in Chapter 1), in that it is probably without much means of testing for the truth of it, at this time. And, most definitely, it is philosophical, rather than scientific. It does aim to provide, a thought-provoking and comprehensive explanation for what is going on – an answer to the bigger question of, "Why are we here?"

The Seven Principles of Love's Agenda

1. Becoming awake/conscious is the primary directive behind evolution, behind all natural processes and lifeforms on our planet

Wherever possible, or opportunity exists, life will emerge, will fight to survive – will be urgent in its intention. While we may consider consciousness – in the form of an awareness of "being self-aware" and manipulation of the environment – to be a bit exclusive to ourselves, as the most potentially awake lifeform, what I am proposing here is that all life is heading the way we are going. In this model, it is not a freak accident, or that we are a freak lifeform, in being conscious. We must not forget either that, even though we are one species, we are already in our billions. We are arguably the most successful lifeform on the planet. Keep in mind our numbers in context with the third part of this theory.

I am here advocating that the "why" of our involution and evolution, or the purpose of life on our planet, is to become awake, become conscious and to contribute to developing wakefulness.

The view I take is that all forms of life, while a part of the bigger ecosystem, are also experiments looking for the best way forward, driven by desire to move from the unawake to being awake. This is Carl Rogers' "actualising tendency" operating towards consciousness. In context there are lifeforms most likely in their design to be able to handle this intention while other lifeforms are less able. But in the mix all provide support to the success of life, and to forms that can obtain consciousness.

In the planetary laboratory, where niche supply and demand operates, there are no guarantees or certainty that lifeforms produced will become conscious or that, if they do, they will not present a threat to each other, and for that matter to the planet itself. Likewise the planet, in seeking to maintain balance, can be a threat to the life it supports – or some of it.

Another angle, on the living Earth/Gaia matter, is to consider that our planet not only supports life but is in the process of conversion – to becoming life or becoming alive. Everything that is born here draws on, and converts, a bit of the Earth to becoming manifest, alive and growing. And life in turn acts as the eyes and ears of the planet to help it further convert to life and adjust to supporting life. Indeed we could say (leaving aside the original Gaia hypothesis), that conscious and semiconscious life is helping the planet itself to become conscious, and that, by this flowering, the Earth comes eventually to know itself, and see itself – through the life it supports.

Let's consider the second principle of Love's Agenda:

2. Becoming awake/conscious is the objective behind all natural processes and lifeforms across the universe

Wherever possible, life will exist and will be urgent in its intention anywhere across the universe. There will be a symbiotic relationship between the given planet and the life forms it sustains, with wakefulness being the primary directive.

By implication we are not alone in this process here on the Earth. I'm suggesting that the ultimate goal of any planet, that can sustain life, is towards a flowering of consciousness. Any planet that can sustain life will be working towards this goal, whether it succeeds or fails, in the short or long term.

Considering the huge number of potential life-sustaining planets, out there in the universe, it stands to reason that there are likely to be planets carrying life that is less awake than life here, while yet others will contain life more awake than life here.

It seems obvious to me that the next huge turning point in our human journey will be in meeting some of the more advanced life, from elsewhere, face to face? I would hope we will meet our Milky Way brothers and sisters in a more convivial open and public manner than the covert meetings that, a number of people believe, have already taken place. See also the Epilogue below.

And so to the third principle of Love's Agenda:

3. All lifeforms are souls. All souls endeavour to migrate towards awakening. They use the "soul ladder" to achieve this

For all lifeforms to have opportunity of becoming conscious, there needs be opportunity for souls operating at even the simplest, rudimentary level of life, at the unconscious reactive level, to travel up through the realms of complexity to the highest level of form and consciousness; that the given host planet can sustain. This opportunity will arise anywhere and everywhere, where life exists. This is evolution (proactive evolution) operating at soul level. I view this situation as two-pronged:

1. All lifeforms have an important role to play in helping to maintain the eco balance of a given planet – and that is why each has emerged there. They have something to do and contribute. Each form will, where needs require, evolve, adapt and refine itself in context with its environmental conditions, and with other species directly linked to it, while meshed in with the refinement intentions of the planet. In context most lifeforms can be restricted or slow in their awakening, with little change over long stretches of time.

2. Any given soul in such limiting circumstances would therefore be severely handicapped with regards to what level of consciousness it could ultimately achieve. That is, if its current form was the only path available to it. But this isn't the route that most will follow. There is, I suggest, a "soul ladder" to awakening. Once one lifeform has broken through the barrier (by its own volition or via help), from the unconscious or semiconscious state, to becoming awake, this opens a path of least resistance, a route, a flow, a stream towards awakening for all souls to follow, from whatever lifeform level they already exist in on that same planet – they are all connected.

Through death and rebirth, there is opportunity for each soul to migrate up the ladder of becoming an ever more complex lifeform towards awakening – to eventually becoming the highest most conscious form the given planet can sustain. In our case it is the human form and the destination of all souls that are involved here on the Earth.

As I see it, this is also a one-way direction only: It would not be in any soul's interest to return down the soul ladder. What I mean is that once a soul takes human form, it cannot go back to a more primitive, less conscious, form. It can only evolve going forward. This view is contrary to what is believed in some Eastern religions, where a soul may return (reincarnate) in a lower, less conscious, lifeform to spend a life.

How a frog becomes a prince or princess

To give an example of how the soul ladder might operate; let me draw on the comment I made in the book's Introduction; regarding a frog becoming a prince or princess.

At the point where we find our frog, it is used to existing as a vertebrate. Following a path of least resistance, the soul of our frog is most likely to follow a vertebrate trail to consciousness. So having spent lives as a tree frog, this soul has advanced enough to move on from being an amphibian, to expressing itself outwardly as a reptile for a number of lives – and takes up being a lizard. At this juncture it has already learnt to balance its body temperatures, and now seeks to learn and experience roaming a wider territory, and being more of a predator along the way.

After a number of lives as a lizard, next it advances again, this time to becoming a mammal in the form of a rodent. Its vibration has speeded up and it is more responsive, more alert – and this time warm-blooded. This experience also helps it to learn about raising and caring for its young. Later in this story, this soul moves up a huge notch and is found spending lives as a pig. It has increased in its size, awareness and intelligence. This expression brings it into contact with humans for the first time, but not necessarily in the best way possible. Later this soul seeks the experience of being a dog, which brings it into closer contact with humans, and on a much better and friendlier footing. This helps it to learn about human ways more intimately.

Much later on in its journey, this soul needs to make a bigger leap. It becomes a primate, living in a forest – a similar environment to the point it started from on this journey, and is happy at that. This time it is away from humans, but has greater awareness and intelligence, and is now in near-human form. Here it is not yet ready to be human but is getting there. Eventually this soul, after probably another thousand years of lives, has moved up to becoming human.

Then, after a series of lives in human form, and becoming more refined, educated and sophisticated, the soul is born into a principality, and becomes a prince or princess; for a life and, who knows, maybe many lives.

Too far-fetched would you say; too much of a fairy story? Well, becoming a prince or princess might be, but becoming human from a frog? As yet, of course, it is a hypothesis based upon belief that can't really be proven one way or the other at this time. But it is not without reason once the "soul" is brought into the equation. You might say that shape-shifting in external expression is inevitably part of the deal.

But is it impossible? I suggest that the answer is a firm No. We need to get past taking everything around us for granted. Given, every lifeform, we now see around us, originated from chemistry, the LUCA and bacteria, then one has to accept that this metamorphosis of life, into its myriad forms, has already happened (and is still happening) on a physical level. I'm simply proposing that it is also happening on an invisible level, with the evolution of souls. Souls, and the cells of organisms, move in mysterious ways… Well it is mysterious – even more so when we fail to keep an open mind about it.

Look around you. We live in a wonderland of lifeforms. Truly anything is possible. And what I'm proposing isn't just an "off the wall" hypothesis; rather I argue that it is vital to making sense, spiritual sense that is, for why we are "all" here. Being in a physical world, with our limited senses, however, it is easier to overlook possibilities and make an argument based upon what is observable, easily measurable and tangible, rather than invoke the invisible for explanation.

And so to the fourth, and an extended principle of Love's Agenda:

4. No stone will be left unturned, and where a planet, or physical body, cannot produce or contain life, to get the conversion going, life will visit it, from elsewhere, and start the process.

Think on the following words of Jesus:

> *I am the light which is over everything*
> *I am the All; from me the All has gone forth,*
> *And to me the All has returned.*
> *Split wood, I am there.*
> *Lift up the stone, and you will find me there.*
>
> Thomas (77)[1]

I believe all the clues are in these words. My contention is that behind this process of conversion there is urgency, a purpose for the "All" to return and therefore no stone can be left unturned in heading towards awakening, to consciousness. If need be, conscious life from elsewhere will eventually visit the seemingly dead worlds and kick-start the life process. Indeed we are already considering doing this ourselves – on the planet Mars. We call it "terraforming."

This fourth principle leads by implication to contemplating a fifth principle (or observation) on Love's Agenda:

5. All matter is seeking to become awake

Extend the theory further and one is obliged to consider that all matter anywhere has a built-in intention and objective to become conscious. If we accept this, then it is logical that even the very atoms that make up everything, are intended to head that way if the journey is at all possible.

> Concerning matter, we have all been wrong. What we have called matter is energy, whose vibration has been so lowered as to be perceptible to the senses. There is no matter.
>
> Albert Einstein

To consider this proposal it is perhaps easier to remind ourselves of Einstein's assertion that all matter is actually energy – a point also made earlier. In this context all energy is seeking to become awake. Rocks operate at a very low vibration, hardly showing a glimmer. This is a slow process too because matter also contains resistance. One could say that it enjoys being matter, or operating at a sluggish vibration and frequency. Indeed, all physical things enjoy their separateness and do not give up their nature so easily. There is therefore a natural resistance to the growth of consciousness and most obviously at this level.

Eventually even the planets, stars, and all forms of energy, will need to somehow be brought into the equation of seeking awakening.

Now to the sixth principle of Love's Agenda, this is another big step:

6. All life and matter, arriving at a state of awakening, of consciousness, needs next to return to Love

Behind all matter and life, and all souls, is Love that is the driver, seeking to be unlocked, recognised, to grow in consciousness and then make the return journey home back to its source. The invisible, through involution, becomes visible and then, through evolution, returns to the invisible. It is Love that is ultimately participating in this process, seeking consciousness and freedom to a higher vibration.

Although from our human level, I believe, we souls make the journey to what we call the Other Side each time we die in our karmic cycle – and then return – ultimately we have to awaken to the point where we leave the cycle, to return to Love. At that point we are without need of returning to the physical.

In the bigger scheme of things, I suggest it is consciousness leading to Love that will also halt the expansion of the Universe and draw it back in on itself to its source.

And there is a seventh principle to Love's Agenda, which is a voluntary yet dangerous step.

7. A voluntary return to the physical

And finally, for those souls who are awake and have succeeded in controlling their involvement in the physical world, they may volunteer to return, in service of aiding all life to become conscious and speed up its return to Love. This is a real test of commitment to Love.

This service may be carried through in many different ways and levels that support the Love agenda and directive. It is not for the faint-hearted however. It is a risky decision, as there are dangers and pitfalls. The challenges faced by returning to physical form, on a given planet, may compromise the spiritual freedom one has already obtained. One may be drawn and, at least to some degree, come to identify with the form one manifests in; and become trapped again in the samsara cycle of that form, and chosen planet.

Recapping

If, having read through the above, you are now contemplating a therapist you could recommend I get in touch with, well let me just take this a bit further by telling you Love's Story, in the next chapter, and see how you view what I'm proposing then.

But let me first recap a little here. In my belief, matter, animals and plants are heading in the same direction as us – towards awakening, towards consciousness. Because of their nature or design, most lifeforms, on the Earth, and as likely elsewhere, do not so easily lend themselves to becoming awake, nor, if they do, by any rapid progress. Some, such as insects, are unlikely to ever get there in the form they are in. They can however get there via the soul pathway I describe. I hold with a view that it is possible for all souls, through death and rebirth, to migrate up, what I call, the "soul ladder" to shift up slowly or speedily notch by notch, to become increasingly more complex and awake as to be able to eventually take on human form – and then onwards back to the source, which is Love.

There is a great little story coming up. Don't miss it...

Notes & references

[1] This version of Thomas found in Sohl, R., and Carr, A. (1970) *The Gospel According to Zen*. Mentor Books.

Love's Story Of Why We Are Here

Chapter 8

Love's Story

Remember to look at the stars, and not down at your feet. Try to make sense of what you see and hold onto that childlike wonder, about what makes the universe exist.

Stephen Hawking[1]

To help illustrate Love's Agenda let me, as Max Bygraves used to say, "tell you a story."

This is the story about Love and written for the child in each of us. The original version of this story took a huge amount of time to write (in a great many parts of the cosmos it is still being written), and to relay, but I negotiated and got access to the shortened version – which saves you spending more than one life reading it. Much of it was written to sound familiar, like it happened in our neck of the woods, and the same for someone else, in their neck of the woods, but it really is a universal story.

Let's begin...

In the Beginning

There was once in the centre of the all-encompassing-darkness a pinpoint of light that knew itself as **Love**.

Now, it could have called itself something like God, Allah, Yahweh, Wakan Tanka, Brahma or perhaps Great Spirit, but it preferred a name that all might accept and feel at ease with, and, importantly, not to be afraid of.

This Love was like an eternal flame that could never die.

As the story goes, Love could not contain itself in this void, this vastness of nothing. It wasn't so much that it felt alone in the darkness but rather its very nature was to give out, to bring some light to the situation. And, of course, there might just be something out there too, that it could give itself to. Yes, well let's be clear about this, Love was indeed also looking for something to love; and there was always the possibility that love would be returned – from, well who knows where, who or what.

But this need for the return of love was a lot more serious than Love made it seem. For if Love spread itself far and wide without love, or enough love, being returned, its power to give love might be seriously compromised.

Anyhow, Love was so excited with the possibilities that it wanted to get on and embrace the void that was all around it. And so it began to give out. It expanded and expanded. Initially it found great joy in this movement, in this reaching out. It was very natural to it, and it did so for a very long time.

In all this movement though nothing was found to love, and certainly nothing was coming back. But Love being ever joyful and unable to contain itself, only increased its efforts and began sending out all the joy and light it could now muster, spreading far and wide into the vastness of nothing.

And still, after a very long time, nothing was found and nothing was coming back...

But something was changing...

What was beginning to happen was that the joy and light that had moved away from the heart of Love, was beginning to cool and slow down a little. And, with this cooling, the joy and light then divided and transmuted into smaller packets of this special light. Of course these were packets of Love, replicas of itself, like globules of pure light. Love well knew what was going on and called the packets of special light, the **Souls**. How familiar that sounds.

Some of the Souls stayed well within reach of the heart of Love and formed a kind of boundary, an intermediary ring-pass-not of light – or so it would seem. As we shall hear these Souls had a special purpose, indeed they were otherwise known as the **Soldiers of Love**. Some of the **Things**, that we shall also hear about shortly, likened the Soldiers to angels, to devas, or higher beings. Although every flicker of pure light was indeed a Soul, the Soldiers were Souls who had attained a higher and finer vibration resulting from their past wakefulness. It only seemed like they were more important than other Souls, when they were not.

The Realm of Things

All was indeed well thus far but as Love poured more of itself into the ether, in all directions so the division that was occurring continued and the cooling also continued.

Now in the furthest nether reaches, of the cold all-encompassing-darkness, a great deal of the pure light had travelled well beyond the boundary of the Soldiers. And there was by now so much of Love out there that it couldn't help but form into beautiful clouds.

If you had gone there, back then, you would have found the void a very misty place to visit indeed. The pure light had slowed to a much slower vibration, and cooled down so much that it became cold, I mean really cold, indeed so cold that the special light was now transforming into something else...

As the clouds condensed and cleared a little, something remarkable was happening! The pure light was becoming clothed, hidden and turning into what we otherwise know as matter. Thus began the process of separation from Love and the beginning of what became known as **The Realm of Things***.*

Things emerged. They came in all shapes and sizes. Some Things were so small that they couldn't easily be seen, some so large that they carried other Things on them. Some Things were as big as planets – indeed they were planets. And some Things were huge and shone like stars – indeed they were stars – and, even greater, they formed into galaxies of Things; a remarkable sight to behold.

　　　　　　Love's Story Of Why We Are Here

The stars and planets, without knowing it, were to play a vital role in this unfolding story. In the world of Things the stars behaved a bit like Love, bringing physical light to all the Things in their keep.

Now let me tell you, as scary as it was to pass beyond the boundary demarked by the Soldiers, this shift into the realm of Things was secretly expected and intended by Love. Indeed Love saw this event coming and embraced it.

You can perhaps guess why...

Yes, Love was now finding something to give itself to. Love sought to embrace the Things and it did so with all the joy it could muster. Ah but if only the Things were so simple as to notice what they were receiving, and what they might be able to offer in return.

Still nothing was coming back to Love... Not even a trickle.

Even so Love was ever prepared to take the risk, and trusted that the Things would change. And fairly, the Things were not really to blame for being unaware of Love. The special light inside all of them was so imprisoned that it was hard to get to. Simply put, the Things had forgotten they were a part of Love. Indeed although they were not yet aware of it, they appeared to be quite happy being Things and separate from each other. It seemed like they were enjoying their freedom and playing in the ether – even though they had no idea of what they were up to.

We need to remember that each Thing had a spark of that eternal flame to keep it going, and so it would happily continue as a Thing, that transformed and was reborn, forever, if left alone.

The Call to the All

This situation went on for a very very long time, with Love giving out and little or nothing coming back. Then a moment came when Love had almost stretched itself to the furthest reaches, and a natural tension arose. There then followed a pause, call it a pregnant pause, where for some time nothing seemed to be happening. All became as if inertia had set in.

*But then, through the use of sound and the flow of, what was described to me as, special music, Love sent out a call. I'm told this is known as **The Call to the All**. It meant it was now time for all of this light and wonderment, indeed all Things, to return to their natural source, back to oneness, back to Love.*

Most Things though, absorbed in being Things, did not hear this call. Fairly they couldn't hear too well, and probably if they could have heard, and think about it, they would have had no intention of returning – and that certainly transpired to be true for some of the Things later. You see they so enjoyed being in the darkness and mostly hidden… It was however vital now that all returned home to that one point.

Stirring and bumping

Many eons passed and the situation seemed at an endless standstill, an endless stalemate indeed. However from inside this long pause, a remarkable event happened. Let me tell you about it, as it was told to me... It sounds truly magical.

A few of the very small, really tiny Things were, deep down, unhappy being alone and away from... Well, they couldn't remember what they were away from. And even though their special light was locked away and not at all awake nevertheless they had a hint, an inkling of a call to action – let's say an inner spark or yearning, a desideratum even.

They began to stir. This is what is remarkable! And in their stirring they became a little bit agitated and animated. They began bumping into each other – very slowly at first. And in this bumping they created something new. They created new Things! It was miracle to behold. The new Things also carried the special light of course, as nothing can exist without it. That's also remarkable! But then the new Things in coming together began forming into small, at first insignificant, groups. Slowly but surely their numbers increased as more and more of them stirred and bumped into each other. All this stirring and bumping caused a chain reaction and their coming together now grew and grew into larger groups.

A good while later in this process and the animated Things were now "two a penny" and they began to become more organised. Their efforts helped their special light to grow brighter and it started to peer through their shells. They were changing, evolving, transforming into new shapes. Some groups became as one organised Thing and that helped them to better deal with their immediate circumstances. They began to bump into each other a little less, and interact with each other a little more.

And it has to be said, that not all was a rose garden in these encounters. Things got lost or destroyed – temporarily at least. We must not forget that they were still not very aware of what they were getting into, and after all they were still Things. But by now they were seeking something – to survive and thrive perhaps. Yes that was enough for now...

Special light shared

All these individual groups of Things were by now producing and reproducing more Things of their own kind, and each time a special light (a Soul) was involved. The older Things died off and released their special light, but each special light, rather than taking the opportunity to return home to Love – as they had been requested to do – could only remember their experience of being a Thing, and so they remained where they were, and became Things again.

Then more sophisticated Things began developing. These were bigger and they could move more easily, see and hear – how truly remarkable is that! They hunted and fed on the less sophisticated Things. They grew bigger and stronger and diversified. There was also now more thought going on in some Things.

And now they became so numerous, there were Things in the waters, Things on the land and Things in the air. And let me tell you this was happening wherever the opportunity arose, particular on all the bigger Things the size of planets, in all the cold nether regions, in every direction. It was a wonder to behold...

Soldiers of Love to the rescue

Well on seeing this effort of Things to stir and awake, and needing urgently to get them heading home, Love decided upon playing a risky move, to speed the process along...

Those small packets of Love that had remained as bright as buttons, the army of Soldiers, were asked to help out. What they were being asked to do now was to volunteer to go to the nether regions, get involved with the Things, and indeed become Things, temporarily at least. Especially to become those that are moving, or growing, and thereby help them to remember where they came from, to help awaken them and steer them back home, back to Love. The reason it was risky of course was because much flow and fluidity had gone out of the equation, out there, and most Things, as we know, were rock solid – in form, substance and disconnection.

And also a great many of the animated Things had indeed, through their experience and learning, organised themselves around remaining in the darkness, and were very protective of their separateness, and the boundaries they had erected to maintain that state. They had fought great battles with each other, over boundaries, to either control the separateness of others, or to protect their own separateness.

They were not intending to return home, not now, not ever. They were happy where they were, and they would certainly put up resistance to any demands placed upon them by this Love. I mean, by Jove, I was told that some of the Things didn't even believe Love existed; they were in such ignorance of the light that resided within them.

It was considered risky for the Soldiers, in carrying out Love's request, that they themselves might also get trapped, by reducing their vibration and becoming Things. If they weren't careful, they could end up being Things again, like everyone else. You see they had experience of this situation before. They knew there was a chance they could get manipulated, and have difficulty remembering who they were, or are. There was indeed a chance that should large numbers of the Soldiers become forgetful of their remit then Love may not have the strength to embrace the All and bring it back to the source – to that one point of light in the void. If this were to happen then the darkness, with or without intention, could claim and trap, perhaps forever, some of the light of Love.

Soul battles

And so it was that the army of Soldiers went out to visit every Thing that could move, or grow, or be showing any signs of awakening. Indeed, it might surprise you, that this also included some of the atoms and molecules. They immersed themselves into Things. Their endeavour, as told to me, was to awaken and release the special light in every Thing and help reverse the situation – to bring all matter back to remembering, back to the light, back to Soul, back to Love.

They met huge resistance and huge battles took place between the Soldiers of Love (who were now also Things of course) and the armies of Things loyal to the darkness, to the separateness.

But greater were the internal struggles, than any external battles. The internal struggles were of the worst kind. It was harsh out there, and in there too. As was predicted, anticipated, some Soldiers, while following their remit, did get trapped, get lost, get turned towards the separation, and became starved of Love – even fought for the darkness. The struggle was long and hard. Nothing was easily gained.

Special training and special Soldiers

Most Soldiers were prepared for what they were getting into. They had their own past experience, and they also relied on others who had more regularly visited the nether regions, by their own volition, and so knew what to expect.

The Soldiers received special training in other words. They learned how not to get trapped, how to free themselves if they did get trapped, when getting involved in Things.

They learnt how to do this through staying for a short time in and around Things, by leaving and returning, and continuing with the work at hand. They learnt through experience, by working with the known laws of Things, by putting right their mistakes when they made them, and, it has to be said, through daily reminding themselves of their commitment to Love. In this, of course, they also reminded the Things around them about Love – which was their main remit.

In some places the Soldiers really struggled to hold onto their link with Love and when and where it got so bad, Love managed to send in a more experienced Soldier (or sometimes a group) who had been around the block a few times, who knew the ropes, to go on a special mission and repoint the way home...

Now I should just mention that you don't get too many of these more experienced Soldiers to the pound, and although some of these very special Soldiers also have the power of magic to impress, and, if need be, to actually coerce Things to head back home, Love had expressly forbidden them to use their magic. In other words to only use the power of love. Love had stressed that all Things must return under their own volition, by their own choice. It was the Law of Love. So things were not made easy for either side.

Love conquers the All

After what seemed like an eternity, it was noticeable that some of the Things were indeed awakening. Some began signing up to want to become Soldiers of Love – and more, many more, were beginning to follow the call of Love. What a remarkable breakthrough that was. This was the best evidence that the Things were on the turn. Love began to hear the voice of Things awakening, of giving back to Love, and calling on their special light to help guide them on their way home. You can imagine this moment was awesome, and moving. Love was now receiving love and was enthralled at this; and indeed embraced the Things all the more.

All the hurt and pain of the separation and wounding was now being understood, now being healed and forgiven – this was unconditional forgiveness as is in the nature of Love. The whole response gained new momentum and Love was so excited and joyful in its embrace – that the All was now returning to its source.

With this movement in full flow, the really big Things, the stars and planets were acknowledged by Love for all the hard work they had done in lighting the void and helping the smaller things to awaken and return to Love. Without them knowing it, it was they, the stars and planets that had fuelled this return and made the whole process possible. The records have it that all the families of stars and planets were "over the Moon" with this praise, and in one joyful moment they each gladly returned to clouds and thus began their homeward journey...

Well to cut a very long story short, Love did eventually conquer the All, by love, and everything returned to its source, as was written. Love threw a party and there was much merriment and great and lasting celebration. Happiness was everywhere for aeons.

All Things come to pass

What a story. What you also need to know with this story is that it was, after all, at that very moment of joy that something else happened – something that we Things all take so much for granted, and it could so easily have been overlooked in the merriment; and in the story for that matter.

With everything returning to that pinpoint of light, Love had actually completed a return breath... Yes a RETURN BREATH. I nearly fell over backwards when I heard this.

Would you credit it? In all of the drama of this story I've unfolded for you, Love had breathed out, giving birth to Things, and had now breathed back in again, completing one full cycle of breath. "Oh, how good it feels to breathe," thought Love.

And so now here, at that very poignant mid-breath moment of peace and completion, Love wondered, albeit briefly... **"How will the next breath go...?"**

The End

And a new out-breath already begun... and WE are in it.

It's all true...

I hope you enjoyed that – and also that in its simplicity it gave you some food for thought in relation to the Love's Agenda chapter.

While I'm still in "story teller" mode, let me share with you that there are some gaps in the story – which is hardly surprising with the aeons it took to run its course. I must admit I'm left wondering whether or not everything did return, or whether there was still some flotsam and jetsam left behind. I think there might be.

But I do need to clarify something in the story that could be a bit misconstrued. Although, in how this story unfolds, it might appear that physical battles took place between the armies of Things and the Soldiers of Love; this was not really the case. You see although the Things did have battles and come to physical blows with each other, across the cosmos, and for long periods, the Soldiers arrival really marked the beginning of resolving the internal struggles that the Things were facing in their awakening. This was the struggle with the resistance they were expected to meet, or had already met, originating from within their own nature.

These then were the battles for their hearts and minds. These were the Soul battles that had to be won from within. For the Soldiers, in becoming Things, once again faced the same struggle against the nature they had now acquired. In this they had to retain their commitment to Love and help to lead the Things out of their own darkness of separation. These struggles were indeed more intense than external battles, yet also subtle and obscure, and not easy to explain unless helped along by a little mythology…

> When I use the word mythology, or myth, I don't mean something that is false. I mean an idea or image by which people make sense of the world.
>
> Alan Watts[2]

By the way… If you thought that the story bore a slightly familiar ring to it, well you are right. Some of this story ties in with our Big Bang theory. Given the story is true, it looks, therefore, as though we really are on the right track with the view that a Big Bang did happen. But it was probably more like a big light show, with a whole lot of Love being expressed. Also, and perhaps more importantly, the story is reminiscent of the Hindu belief involving the *Days and Nights of Brahma*. Indeed in part it is very similar. And that's remarkable! It proves those Hindus knew a thing or two, even all those thousands of years ago. As Brahma breathes out, all things manifest, and as he breathes back in, all things return to their source.

Mind you, those Hindus may not have been right about all things. They have always seen this bigger picture as an ongoing cycle of birth, death, rebirth, and re-death, ad infinitum – or so it seems. The intended impression one gets from the Love Story though is that, rather than an endless repetitive cycle, it is about Love, in its experiencing, also being on a learning curve. With each breath, Love opens to new experience on its journey towards its own enlightenment, to becoming more awake. Love opens to yet new adventures and learning. So yes, put another way, Love is learning how to be more loving, Brahma is learning about being Brahma, God is learning about being God. And each soul is learning about becoming a vital drop of Love – and awake.

I also noticed something else in the story, after frantically copying it down. This is regarding love for oneself. If you read between the lines, Love was seeking to love itself. After all, the Things were not as they seemed were they? They were indeed already a part of Love but not awake to that, and neither it seemed was Love entirely awake to acknowledging that either. You could say the whole caboodle was an illusion, but very real, nevertheless. I have a suspicion that Love knew this, and that this element in the story was meant to be discovered by contemplation. If that's the case, I've unfortunately blown the cover on it.

Why we are here

If you get it, the story provides a simple cosmology, a framework that is intended to help make sense for why we souls have either come to this planet, or developed here. It is why we are here now. And it also provides a working hypothesis for our bigger purpose here and elsewhere in the universe.

So what convinces me that this story could be true? Well, as has been a theme running through my first book,[3] it is making sense of circumstances we all live and have our being in. It is the simple yet remarkable fact that we are here – along with the rest of the flora and fauna – that convinces me. I believe too that there is this process of refinement going on in the flora and fauna, so that the plants and animals we live with today, in nature, are more refined than they would have been in antiquity – obvious examples would be dinosaurs and hominids then, in context with birds and us humans now. As I've argued, evolution, as currently posited doesn't account for any tendency towards refinement, meeting objectives or direction. The process can go either way in response to the environment. I obviously believe there is a plan towards refinement, and a purpose behind that plan.

> According to this hypothesis, as developed during the twentieth century by Bergson, Bucke, Julian Huxley, Teilhard [de Chardin], Jung, Medawar, and others, the main thrust of evolution is to develop an increasing capacity for breadth and depth of awareness, with the multiplicity of physical forms a mere by-product of this central evolutionary drive.
>
> Arthur Ford[4]

My belief then is that we are all working towards becoming awake – ultimately to return to Love. Clearly from the Arthur Ford quote above I'm in good company in this belief – at least in part. How pleased I was, to discover this little gem of a quote supporting the general hypothesis. To my mind, of course it's not much of a further step to suggest that what we see as non-biological inanimate matter, that makes up the bigger part of what is around us, is also, painfully slowly, going in that direction too. Not to labour points made earlier: We are a part of this planet becoming conscious of itself; and further, a part of the universe becoming conscious of itself.

In our evolving wakefulness, we humans are not easy to explain away other than being here for a higher purpose. The Earth clearly is more conducive to the growth of consciousness than other planetary bodies around – particularly where there is no life as we know it as yet, or where life has tried and so far failed. But "no stone will be left unturned" takes on a different meaning when considered in context with such a universal intention.

And what can we do about it all? I trust the next chapter may provide food for thought.

Notes & references

[1] Hawking, S. (2013, Nov 18) *Stephen Hawking on black holes*. The Guardian, on YouTube [Accessed 03/04/2018]

[2] Watts, A. (2011 edition) *Eastern Wisdom, Modern Life*: Collected Talks 1960–1980, p. 99.

[3] O'Neill, F (2016) *Life and Death: Making Sense of It*. Some Inspiration Publications.

[4] Ford, A. (1974) *The Life Beyond Death*. Abacus.

Chapter 9

Remembering we are Soldiers of Love

So now let us bring Love's story back to us, living on this planet...

As souls (spirits) let me suggest that we entered into a symbiotic relationship with our human bodies from the time of our conception. None of us is a tabula rasa at the get-go. Each of us was born at a significant moment in time, tying in with what we each need to experience, learn or express in our lifetime. Essentially, however which way we cut it; this is to become more loving beings. But in what way, in what experience and what kind of giving, we must, of course, discover this for ourselves. There is nothing fixed other than in symbolical terms, and the conditions we face with our start in life may support or tend to hamper, or cause resistance to our success.

You and I are here to love and respect each other and help the rest of the flora and fauna, where possible, to flourish. We all know this in our hearts of course. We learn to do this through our kindness towards all life, our interaction with our environment, and exchange of energy fields – through the love we give out and receive – creating healthy karma and bringing goodness to life, and lives in our care.

If we have grasped Love's story (and, in keeping with it), we will also know that we *Things,* as we are now, need to be on the turn and heading homeward – the *Call to the All,* went out aeons ago. To achieve this, it will of course help if we all remember who we are. Some of us may well already remember we are the *Soldiers* (of the story), but for most of us it is probably not an easy way back to self-remembering.

How we get back to being Soldiers of Love

As you will probably know, a number of the "not too many to the pound," special *Soldiers of Love* have had to visit our planet to set us on the right course – yes things have been that desperate at times. And the wisdom they left can be found if we search for it. On that note I can help to give you a head start in your search. I've been digging around for such wisdom for some time; and I've found a few nuggets they have left behind. Here are some, let's call them "conditions," that you or I will need to meet, get our heads around, in order to make such a change of direction to happen:

Look after yourself

This one is obvious, but we don't all pay heed to our bodies. Each of us needs to eat healthily, sleep well and take regular exercise. We need to keep our body as fit and as healthy as we can. Get inspired too, and get some laughter into our lives. Spend time with others. Music, nature and comedy can be great life-invigorating, life-enhancing and life-giving devices. Also we need to take a risk or two in life – and not linger for too long, at any time, in any "comfort zone" we have created.

But importantly, this is also about diet of the mind as well as the body. We are what we take in, what we eat, drink, think, see, hear and say. It's about mind, body and soul being aligned – having all our ducks in a row.

Be mindful of one's commitment to Love

To get back to being a *Soldier of Love* we need to be ever mindful of our commitment to Love – Love's story made that clear. We will need to get with, and stay with, Love's Agenda, no matter how hard it gets, and how much temptation, to veer away from that commitment, is placed in our way – and there will probably be many trials that we face that test that commitment.

Be self-sufficient

We may well falter at times. We may well feel alone and isolated at times. We are away from home. This is why it is also important to daily remind ourselves of both our commitment to Love, and to become self-sufficient. We will certainly be loved in return, although it may not manifest in ways we might expect. There could be times indeed when we may feel we have been totally abandoned. The *dark night of the soul*, is a well-known state to the *Soldiers of Love*.

Be mindful in relationships

Does it need saying how important it is to let those we love know how much we love them? I doubt it, but probably we don't say it as often as we might. Equally it is important to resolve any disagreements and disharmony in our relationships. Don't let hatred, envy or jealousy smoulder inside. We need to leave resentments and regrets behind on our journeys. It lessens the load, the baggage we are carrying – travel light.

Practice meditation

Regular meditation will help to calm us down, quieten the mind, and help to keep our heads clear. We need to make it a lifestyle choice. Get into the habit. This is crucial practice to any routine living, and importantly to open the heart and head to other possibilities. See below for access to some simple tips on meditation.

> *Learn to be empty*
> *Of all things, interiorly*
> *And exteriorly, and you*
> *Will behold your God*

John of the Cross[1]

Practice contemplation

Study the deep and profound, and especially draw on your own treasure house.

> Daiju visited the master Baso in China. Baso asked: "*What do you seek?*"
> "*Enlightenment,*" replied Daiju.
> "*You have your own treasure house. Why do you search outside?*" Baso asked.
> Daiju enquired: "*Where is my treasure house?*"
> Baso answered: "*What you are asking [with] is your treasure house.*"
> Daiju was enlightened! Ever after he urged his friends: "*Open your own treasure house and use those treasures.*"[2]

Love's Story Of Why We Are Here

While as a *Soldier of Love* our remit is to help awaken life. This cannot truly take place until we ourselves are self-remembering and awake.

It should probably go without saying that we, each of us, need to ask, and find answers to, not only the bigger questions life puts before us – such as, is there a God; is there a life after death; have we lived before; why are we here; and is there intelligent life out there – but firstly to ask questions about ourselves. Questions around why I am here; what is my purpose; what could I be seeking to learn (or give) about myself; what is the symbolism and meaning behind where I grew up and the people I grew up with. You and I will be here for a reason, a meaningful purpose that needs to be solved, uncovered and acted upon. It is part of the wonderful adventure we are on.

If not already doing so, we each likewise would be encouraged to get studying philosophy, psychology and spiritual texts. Dig into any such knowledge or books that appeal. For example, look into the standards such as The Bible, The Koran, The Tora, the Tao Te Ching, or The Tibetan Book of the Dead. And possibly read books that help lift us towards loftier heights in our knowledge and understanding of human behaviour and spirituality. I'm thinking of such authors as C G Jung, Sogyal Rinpoche, Pierre Teilhard de Chardin, Carl Rogers, Alice Bailey, Stephen Covey, Charles Haanel, Napoleon Hill or Robert Fritz. But really whoever rocks your spiritual boat.

Also we'd be well advised to delve into an holistic art, such as astrology, the tarot or numerology to, at very least, get a sense of how they work. I guarantee you'll be surprised how much wisdom these arts carry when studied correctly. We might also check out the Ten Bulls of Zen to help us identify our goals and gauge where we may be now in our spiritual endeavours. My own plea here: Most definitely explore using the I Ching to help answer profound questions you have; to uncover truths and provide guidance along your path.

Practice gratitude and compassion

It is important we find gratitude, and forgiveness, in our hearts – for ourselves as much as others. This helps us to hold our balance and be in the present without baggage. Being grateful, giving and loving to all people, creatures and plants also feeds back to us, nourishing the soul.

The Buddhist Eightfold Path is to be recommended for the way to live with gratitude and compassion. Siddhartha Gautama is attributed with originating the idea of liberation from the cycle of birth and death. He posited the Eightfold Path to freedom, in order to achieve this. This path is based upon following a simple eight point formula: Right view; Right intention; Right speech; Right action; Right livelihood; Right effort; Right mindfulness; and Right concentration. See below for where to get more information on this path.

Be lofty, urgent and of service

The general recommendation is to keep one's thoughts and energies up there towards love, peace and goodwill to all.

It is important also, when the time is right, to find some means of service to others, in a way that may help change their lives for the better – and hopefully help get them heading "home" to Love. It would require we set our lives going from here on in as though we only had a short time to live. Important things require a touch of urgency in other words.

Be mindful of death

And it follows. To be the *Soldiers of Love* we are, we need to be ever mindful of death, or in other words to be awake to the transitory nature of the situation we find ourselves in, on the Earth. This is not to forsake being cheerful and ever in good humour. The vibration we are at now, is only temporary, a small part of our journey home. It is, as I'm sure you will know a lower vibration and frequency than our more natural state. We need to be careful then to learn from our experience here but keep mindful of not getting over-involved in it. All is transitory and will pass eventually.

Comment

Putting these conditions into place is only a part of the practice of being a *Soldier of Love*. We have to want to do it and remain disciplined in application. As implied in Love's Story, it is not easy to make and retain such a commitment. Let's not forget, we're running against the tide of our animal nature.

A helpful resource

If it interests you, my book, *Steps to Health, Wealth & Inner Peace* supports good practice with some of the concerns and conditions described above – notably on meditation, practicing compassion, and The Eightfold Path. It is available through most online stores in ebook or paperback.[3]

Note on making a physical contribution

By just being in this physical experience our soul influences the very cells, indeed the very elements and atoms that make up our body. We are capable of lifting or lowering the quality of this influence, and we are encouraged, through our thoughts and actions, to help the whole structure to become more awake, more alive. This, I believe, has a knock-on effect for our health, and the planet.

Our body will be exchanging dead cells for new ones throughout our lives. Eventually the body will die and be left to either decompose or be burnt; and the informed "CHNOPS" (elements of carbon, hydrogen, nitrogen, oxygen, phosphorus, and sulphur), will be released to be recycled back into the atmosphere, back into the flora and fauna for new life.

Is there is a chance that memory is carried with these elements? If so, and if done right, all the goodness, we created and retained through life, will contribute to the continued growth of life and, thereby, support life awakening, becoming conscious. Well it's a thought...

Final thought

Perhaps of equal, or bigger concern that this book needs to get across, is that the next time you are having a duel with a mosquito, or a wasp, think on. They too are souls, like you, finding their way to awakening...

On that note, I'll buzz off... Have a good journey home.

Affairs are now soul size.
The enterprise is exploration into God.
Where are you making for? It takes
So many thousand years to wake...
But will you wake, for pity's sake?

Extract from *A Sleep of Prisoners* by Christopher Fry[4]

Notes & references

[1] Campbell, C. A. (1989) *Meditations with John of The Cross.* Bear & Company.

[2] Sohl, R., and Carr, A. (1970) *The Gospel According to Zen.* Mentor Books.

[3] O'Neill, F. (2016) *Steps to Health, Wealth & Inner Peace.* Some Inspiration Publications. Contains a lot of the advice given here. You can get it from most online stores.

[4] Fry, C. (1951) *A Sleep of Prisoners.* Extract from the poem/play.

Epilogue

Change is a-coming: Awakening and Disclosure

I briefly want to pick up again on UFOs and their extraterrestrial implications here.

Our human story is inextricably linked, and running parallel with a growing awareness, a growing wakefulness. It represents the rise of consciousness on our planet.

In my *Life and Death* book I muse on possibilities for how our human development came about, in practice that is. From the evidence we have – the artwork mainly in clay and in caves – it was a fairly sudden event by evolutionary standards. For millions of years, it appears, we were pretty much eking out a survival existence, living a nomadic lifestyle, and making use of shelter where we could find it. We had learnt, later, to use fire and make simple tools for hunting prey. In the Life and Consciousness model (see Chapter 6), I would place us at Level Three to Four back then. But almost the next minute there is a leap in awareness, in intelligence and creative application. There is culture arising.

This was all going on somewhere between a hundred thousand and fifty thousand years ago. Against the millions of years before; this time period was a drop in the ocean. It would still be a "drop in the ocean" if pushed back another hundred thousand years. Something remarkable happened. Suddenly, as I see it, we were awake. We had become human beings. One of the suggestions I make for how it may have come about was that we, our "ancestral we," were helped along by extraterrestrials. Our ancestors were visited by a more advanced civilisation – most probably from elsewhere in our Milky Way galaxy. Possibly through genetic modification and/or interbreeding they made changes to the hominid design that we stemmed from. The changes ultimately raised us to new levels as a result.

Tipping point

Come back Erich, all is forgiven...

Now this theory of extraterrestrial involvement, of course, isn't new. Erich von Daniken[1] was making such claims back in 1968, with his take on some of the more bizarre archaeology we find around us. Regardless of how damned, as pseudoscience, his books were by the scientific community (I had my doubts at the time, and I so recall how my fellow archaeologists scoffed at his theories and suggestions), the more I consider it the more I'm inclined to believe that he was/is on the right track. I'm of the belief that we looked the part (bipedal form), were heading in the right direction and getting pretty close to awakening, by our own volition, and that, that extraterrestrial nudge pushed us over the tipping point to become fully human. We were helped along, just as we have done for other species, and no doubt will be doing probably also out there, as here – in the times to come.

Much more recently, Robert Lazar (of Area 51 fame, or "infamy" depending on your level of scepticism of Lazar) made claims that extraterrestrials from *Zeta Reticuli 1* and *2* star systems had been visiting Earth for a long time and that we humans are a *"product of externally corrected evolution ... that Man as a species had been genetically altered sixty five times."*[2] Some food for thought methinks – I believe his account.

On the edge of something big

But with regards to the process of awakening, let me tell you that from my perspective we are on the edge of having to become a lot more awake, and more quickly. Things are speeding up towards a discovery that could be so big that it will blow our socks off. It will require us to shift into another gear in our understanding. This will have a dramatic effect on our constructs, beliefs about who we are and where we come from. What it is to be human. What it is to be an Earthling indeed. It will have impact on our social structures and particularly on our religious beliefs. Holding onto the older paradigm of God in the clouds, with heaven and hell, may well prove to be difficult or painful. Such belief could very possibly contribute, if not actually provide, the cause of our slow acceptance of the incoming paradigm – that I call the *holistic-spiritual paradigm* in the broadest sense.

No doubt a lot of what we are expected to get to grips with will be denied until, that is, we are ready to accept the new game we're playing. So it won't just be in our thoughts and ideas, or ideals for that matter, but in how we feel, at an emotional/gut level, how we belong or where we belong. All our collective primal fears (that we could associate with Maslow's Physiological Needs, Safety/Security Needs, and a Sense of Belonging), will probably resurface to be re-evaluated.

What's on the cards

What we are going to discover, in a profound and self-evident way, is that we, and other lifeforms on the Earth, are not the only lifeforms in the Universe. We are going to discover first-hand that we, and the flora and fauna of the Earth, are not the only intelligent life in the Universe, or the most intelligent. Our biggest shock, I suspect, will be to discover that we have brothers and sisters in other parts of our galaxy, beings that will look similar to us – or vice versa.

Our more technically advanced brothers and sisters have possibly learnt how to live much longer than we do. We can surmise that some will be more conscious and spiritually advanced than we, having travelled that much further in understanding themselves – and have gotten closer to the source, to *Love* as I am calling it. Others may not be so awake – rather they could be technically far more advanced than we but in other respects less developed mentally and emotionally than we are – and trusting them may be questionable.

If you think about the diversity of development, culture, beliefs and values in humanity, across the board, it probably stands to reason there is likely to be equal diversity out there too, and not all of it welcome. Some of, what we may assume to be organic, may in fact be android in nature. It is a path we are also treading.

And let's consider that some of these beings, which we describe as "aliens," could, shock, horror, already be living here beside us... There are experts in the field (Linda Moulton Howe for example) who believe this to be the case. After all the

Earth is a very attractive blue planet, with a great atmosphere, and it should come as no surprise, that other beings have been visiting us – whether to spark their particular kind of life, whether on learning trips, or holiday visits, or to settle here – for a very long time. And what is possibly more profound, some of us could have done the return favour. I'm thinking less in terms of possible abductions here and more in terms of a soul choice to have experienced physical living/learning on habitable planets other than the Earth.

Not that new...

I should flag up at this point that while I'm presenting this as a vision for what is coming, it is not that new an expectation for quite a number of people. If you have kept abreast of developments it is in effect old news. And old news that is yet to be fully taken up by conventional media as serious news. There is a movement, coming from many quarters now, that is requesting "disclosure." This is a call, in other words, for the known facts about UFOs, and indeed extraterrestrials, held by governments, to be put out there into the public domain. We're not just talking about a few space cadets making noises either, no this is for real. For example, visit the Disclosure Project where it is stated:

> We have over 500 government, military, and intelligence community witnesses testifying to their direct, personal, first-hand experience with UFOs, ETs, ET technology, and the cover-up that keeps this information secret.[3]

The hope, and anticipation, is that governments across the developed world will announce, sooner than later, what they know on UFOs and contact.

As yet though, because there is reluctance to release this information, it is only seeping out very slowly into the public domain – or purposely being drip-fed. Consider for example the unclassified, US Government's, *Tic Tac UFO Executive Report*,[4] circulating from 2017, that is considered by UFO commentators/critics as legitimate. The report accounts for a series of incidents involving UFOs (or AAVs) that took place over the Nimitz Carrier Strike Group, in 2004. It accounts for the various advanced sensors being used during the incidents, and the resulting determination that the advanced technology observed was far beyond any known US or international government capability. The report of what happened makes for very interesting reading:

> During the period of approximately 10-16 November 2004, the Nimitz Carrier Strike Group (CSG) was operating off the western coast of the United States in preparation for their deployment to the Arabian Sea. The USS Princetown on several occasions detected multiple Anomalous Aerial Vehicles (AAVs) operating in and around the vicinity of the CSG. The AAVs would descend "very rapidly" from approximately 60,000 feet down to approximately 50 feet in a matter of seconds. They would then hover or stay stationary on the radar for a short time and depart at high velocities and turn rates.

Returning fighter planes, from a training mission, offered opportunity for an attempt to intercept:

> On 14 November after again detecting the AAV, the USS Princeton took the opportunity of having a flight of two F/A-18Fs returning from training mission to further investigate the AAV. The USS Princeton took over control of the F/A-18Fs … for intercept leading to visual contact approximately one mile away from the AAV, which was reported to be "an elongated egg or a 'Tic Tac' shape with a discernible midline horizontal axis." It was "solid white,

smooth, with no edges." It was "uniformly coloured with no nacelles, pylons, or wings." It was approximately 46 feet in length. The F/A-18Fs radar could not obtain a "lock" on the AAV; however it could be tracked while stationary and at slower speeds with the Forward Looking Infrared (FLIR). The AAV did not take any offensive action against the CSG...

The report goes on to say that, "given its ability to operate unchallenged in close vicinity to the CSG it demonstrated the potential to conduct undetected reconnaissance, leave the CSG with a limited ability to detect, track, and/or engage the AAV."

Probably most of us are still unaware of it, or choosing to be unaware. We continue to view the notion of alien visits as almost comic book, bizarre and wacky. It's treated as on a par with Micky Mouse, mad conspiracy theories and "post-truth" tabloid banality.

We have nevertheless been preparing ourselves for this revelation, this disclosure, for some time whether we are fully aware of it or not. This is in part driven by the exposés and conspiracy theories kicking around – such as with "Area 51" in the USA. It is in part driven by what we are increasingly observing in the skies, and recording with our mobile devices; or by what has occasionally been reported in the press and social media.

Probably the biggest driver though is in what we have envisioned in our imagination and created through sci-fi story, movie and drama. The evidence is already mounting for this breakthrough. Governments, and individuals, are coming forward, are slowly letting us know that UFOs are being taken seriously – UFOs are real, and extraterrestrials likewise are real and could be amongst us.

A comment on Crop-Circles

I make no apology for dropping in a comment on crop-circles here – particularly in respect of their possible link with UFOs. Did you know that crop-circles are still being made each year in all sorts of places, and not all are immediately in the public eye?

> The documented cases have substantially increased from the 1970s to current times. Twenty six countries reported approximately 10,000 crop-circles in the last third of the 20th century; 90% of those were located in southern England.[5]

Clearly crop-circle makers were certainly busy in the last century, but it is still going on, just not being so regularly reported in the media. If you haven't already done so, you really ought to take a look into the phenomenon. Some of the ones being found now are very sophisticated, beautifully designed and beautifully executed. Sceptics are adamant that they are all made by human hand, but likewise many people, including myself, doubt it.

Supposedly teams of people are going out each night of the summer months to make them. Isn't it amazing what we can create, and in the dark too? Some of the patterns clearly take great intelligence, skill and co-ordination to make. Whoever is making them, they are doing an amazing job.

Let me say that apart from surprisingly sensible people making claims of light orbs and UFOs being involved, in the creation of some crop-circles at least – with the occasional video being captured to demonstrate it – three considerations make me question that all crop-circles are supposedly of human origin:

The first is the regularity of their occurrence year in year out – particularly in Wiltshire UK. It strikes me to be such a faddy thing to be involved with, at a human level that is. Especially since the cat is supposedly out the bag and we now know (well we are told) only we humans could have made them. I mean, what is the motivation, what is to be gained now by doing another nice crop-circle – and ruining some farmer's wheat crop to boot? What is wrong with having a good night's sleep instead of trying to convince people that the aliens have arrived? Okay so there's a challenge, and competitive crop-circle makers are amongst us, who like to work on a big canvass. And their Wednesday night team, with very simple tools, like to beat the Thursday night team with all their whizzy corn-breaking paraphernalia. Well okay I suppose so... Sorry I still have my doubts.

Here's a second concern. I don't know about you, but the size, beauty and intricacy, of some of the patterns, that are also created in a very short time too, takes my breath away. Clearly some of us busy ourselves with making intricate shapes, curves, triangles and circles in the wheat, to produce a structure that can amount to hundreds of metres in size. We, if it is just us, have even, on occasion, left one or two challenging shapes behind. For example, a Mandelbrot set (1991) in Cambridgeshire; a representation of Pi (2008) in Wiltshire; a message in binary code (2002) in Hampshire; and an answer to the Arecibo message (2001) near Chilbolton radio telescope. The original Arecibo message was sent into space by Frank Drake's SETI team of scientists back in 1974.[6]

And we are doing this at night, in our wellies, with boards, bits of rope, and not forgetting a good eye for detail. Wow, we are incredible. It has been noticed too, on occasion, that relatively small outlying circles, made in the corn, can appear without any means of access – such as the tractor tracks that could normally be used. The only access would be from above. That takes some doing. Maybe the makers are using some kind of pole-vault method to achieve getting in and out. Human ingenuity is amazing but I do wonder…

My third concern, and what sorts the chaff from the wheat literally, is the condition of the wheat (or whichever crop is being disturbed), and also the abnormality of the soil in the immediate area of the crop circle. It is argued by crop-circle researchers that microwave radiation can be traced where the genuine article is involved – and correspondingly not so when made by us. The crop stalks are bent over by sudden heat – and the stalks show evidence of popping, or being blown out, on the bend, due to the heat. Notably in respect of this; in 2002 there was an attempt to show we can do the microwave thing too. This experiment was carried out by students from the Massachusetts Institute of Technology.[7] They tried to reproduce the microwave effect on the wheat, while creating a crop-circle, but in vain.

A number of researchers claim the patterns are intended as communications with us, through intuitive symbolical terms. The binary code and Arecibo answer patterns (mentioned above) could be sighted as obvious examples.

> One theory is that they are trying to introduce themselves to us peacefully, like we do using bubbles with dolphins. When humans want to talk with dolphins we put little bubble circles under the ocean with a generator. We watch the dolphins come up and play and investigate, and we study them. This is called the "dolphin communication project." The dynamic between the way humans interact with dolphins and whales is likely comparable to how extraterrestrials communicate with us.
>
> Dr Horace Drew, molecular biologist[8]

The idea that crop-circles are intended as some form of peaceful communications with us is worth serious consideration. Personally, if I may, I believe we should get scientific about this matter; and entertain serious doubt that all crop-circles are made by us... Thank goodness, some researchers are doing this already.

On that note, I'd like to give a mention to Lucy Pringle, probably the best known photographer and researcher of crop-circles in the UK. She has been photographing them for a number of years. Check out the collection on her website: https://cropcircles.lucypringle.co.uk/

Yes, we really do live in exciting times, don't you think?

Explore more about UFOs and extraterrestrials

If you are interested to find out more about UFOs and extra-terrestrials there are a few useful links to visit below. There are many more to be found on the Web. If you want to get out there, I'd encourage you to take a look at the live feeds from the NASA – find on www.nasa.gov/multimedia/nasatv/. By the way, if you want to see something phenomenal, that has been regularly observed by bystanders and scientists alike, check out the Hessdalen Lights in Norway. These have been seen since the 1930s but really got going in the 1980s. Some people think it is a kind of plasma earth energy, or ionised gas, that we don't yet fully understand. Others claim to have seen cigar-shaped spacecraft associated with the lights phenomenon. It has noticeably quietened down over the years, but there are plenty of records of their appearance. You can view the Hessdalen Lights on YouTube, but also check out the project website.[9]

A few UFO organisations – Europe, Russia and USA*

- BUFORA British UFO Research Association
  https://www.bufora.org.uk/
- CNES Centre National d'Études Spatiales
  http://www.cnes-geipan.fr/
- CUFOS Center for UFO Studies
  https://www.cufos.org/
- DISCLOSURE PROJECT
  http://www.disclosureproject.org/
- MUFON Mutual UFO Network
  http://www.mufon.com/
- MUFON-CES Mufon in Germany
  https://www.mufon-dsr.com/
- NUFORC National UFO Reporting Center
  http://www.nuforc.org/
- RUFORS Russia
  http://www.rufors.ru/
- SUFOI Skandinavisk UFO Information
  http://www.sufoi.dk/
- UFO-Datenbank.DE Germany
  https://www.ufo-datenbank.de/
- UFO Sweden
  https://www.ufo.se/

*Information correct at time of writing.

Notes & references

[1] Von Daniken, E. (1968) Chariots of the Gods. First published in 1968 by Econ-Verlag (Germany) Putnam (USA).

[2] Lazar R, (2010, 20 November) *UFOs & Area 51 - The Official Bob Lazar Video - Alien Technology Revealed.* UFOTV® The Disclosure Movie Network https://www.youtube.com/watch?v=IJolFbj8nc4 [Accessed 21/05/2015]. This is a remake of an earlier film.

[3] Disclosure Project. Run by Dr Steven Greer. Quote taken from the Home page. http://www.disclosureproject.org/ [Accessed 13/02/2018]

[4] The Tic Tac UFO Executive Report is reportedly to have been first uncovered by the Las Vegas News Network (KLAS). It can be readily found on the Web in PDF format. If you need a copy from the author, use the contact link in SomeInspiration.com

[5] Dennis4706 (November 2012) *"So you dont believe in Crop Circles"* (Watch This) YouTube video. https://www.youtube.com/watch?v=CuE7-VYExPg [Accessed 02/05/2017]

[6] The Arecibo message, part of the Ozma Project (involving Dr Frank Drake) to locate extraterrestrial life, was beamed into space a single time at a ceremony to mark the remodelling of the Arecibo radio telescope on 16 November 1974. It was aimed at the globular star cluster M13 some 25,000 light years away. The total broadcast was less than three minutes. To anyone listening, the message they received was in binary and tells facts of our human story. *Crop Circles The Arecibo Reply* provides a good overview of the Arecibo crop-circle. https://www.youtube.com/watch?v=D1QWXjLtA84 [Accessed 10/01/ 2017]

[7] Talbott, N. (2002) *M.I.T. Kids' Crop Circle Attempt Yields An Interesting (And Totally Inadvertent) Result.* BLT Research Team Inc. http://www.bltresearch.com/published/mit.php [Accessed 03/04/2017]

[8] Dr Horace Drew comments taken from https://www.news.com.au/technology/science/space/theyre-real-and-contain-hidden-messages-scientist-says/news-story/05f450e958f71dac46a2a61f47968f16 [Accessed 20/06/2018]

[9] Project Hessdalen http://www.hessdalen.org/index_e.shtml [Accessed 09/12/2017]

Bibliography

Books

Baba, M. (1967) *Discourses 2 San Francisco: Sufism reoriented* P145. ISBN 978-1880619094

Brockman, J (1995) *The Third Culture: Beyond the Scientific Revolution.* Simon & Schuster.

Campbell, C. A. (1989) *Meditations with John of The Cross.* Bear & Company.

Capra, F. (1983) *The Turning Point.* Flamingo/Fontana Paperbacks. P76.

Carson, R. (1962) *Silent Spring.* Houghton Mifflin. P297

Darwin, C. (1859), *The Origin of Species by Means of Natural Selection,* John Murray, p. 490.

Ford, A. (1974) *The Life Beyond Death.* Abacus.

Gingerich, O. (1993) *The eye of heaven: Ptolemy, Copernicus, Kepler.* American Institute of Physics, 181. New York.

Haanel, F. C. (2016) *The Master Key System.* Some Inspiration Publications. Item 34 p10. First published in 1916 by Psychology Publishing USA.

Heisenberg, W. (1962) *Physics and Philosophy.* New York: Harper and Row. p.145.

Kirchner, James W. (2002) *The Gaia Hypothesis: Fact, Theory, and Wishful Thinking.* Climatic Change 52: 391-408. Kluwer Academic Publishers.

Lovelock, James E (1979) *Gaia: A New Look at Life on Earth.* Oxford University Press.

Lovelock, J. (1991) *Healing Gaia: Practical Medicine for the Planet.* Harmony Books: New York

Lovelock, J. (2009) *The Vanishing Face of Gaia.* Basic Books.

Moray, N (2014) *Science, Cells and Souls: An Introduction to Human Nature.* AuthorHouse UK.

Newport, John P. (1997) *The New Age Movement and the Biblical Worldview: Conflict and Dialogue.* Wm. B. Eerdmans Publishing Company.

O'Neill, F (2016) *Life and Death: Making Sense of It.* Some Inspiration Publications.

O'Neill, F. (2016) *Steps to Health, Wealth & Inner Peace.* ISBN: 978-0-9934626-3-4. Some Inspiration Publications.

Rogers, C. (1978) *Carl Rogers on Personal Power.* Constable London.

Sohl, R., and Carr, A. (1970) *The Gospel According to Zen.* Mentor Books.

Von Daniken, E. (1968) *Chariots of the Gods.* First published in 1968 by Econ-Verlag# (Germany) Putnam (USA).

Watts, A. (2011 edition) *Eastern Wisdom, Modern Life:* Collected Talks 1960–1980, p. 99.

Articles

Blair, T. (October 4, 2016) *Is Gaia Dead?* The Daily Telegraph.

Eggenstein, K. (1984) *Materialistic Science on the Wrong Track.* http://www.chemtrails-info.de/kee/1/i-mater.htm

Ellefson, L (December 21, 2017) Neil deGrasse Tyson on Ufos: 'Call me when you have a dinner invite from an alien' CNN News http://edition.cnn.com/2017/12/20/us/neil-degrasse-tyson-ufos-new-day-cnntv/index.html

Fry, C. (1951) *A Sleep of Prisoners.* Extract from the poem/play.

Harvard University (2012) *Gaia hypothesis.* Gaia-hypothesis-wikipedia (pdf file).

Hawking, S. (2013, Nov 18) *Stephen Hawking on black holes.* The Guardian, on YouTube

Kleidon, A (Mar 2011) *How does the earth system generate and maintain thermodynamic disequilibrium and what does it imply for the future of the planet?* Article submitted to the Philosophical Transactions of the Royal Society.

Lazar R, (2010, 20 November) *UFOs & Area 51 - The Official Bob Lazar Video - Alien Technology Revealed.* UFOTV® The Disclosure Movie Network https://www.youtube.com/watch?v=IJolFbj8nc4. This is a remake of an earlier film.

Lipton, Dr. B. H. (2000) *Biological Consciousness and the New Medicine.* http://www.dragonway.co/essays/biological_consciousness.htm

Maslow, A. (1943) Paper: *A Theory of Human Motivation.* Psychological Review, 50 (4)

Scudder Dr T. (2012, June 29) *Hidden Portals in Earth's Magnetic Field.* NASA Science Beta. https://science.nasa.gov/science-news/science-at-nasa/2012/29jun_hiddenportals/

Woese, C. R.; Kandler, O.; Wheelis, M. L. (1990). *Towards a natural system of organisms: proposal for the domains Archaea, Bacteria, and Eucarya.* Proceedings of the National Academy of Sciences. 87.

Websites/URLs visited

BBC
http://www.bbc.co.uk/nature/21923937
Daily Mail newspaper
http://www.dailymail.co.uk/sciencetech/article-1037471/Apollo-14-astronaut-claims-aliens-HAVE-contact--covered-60-years.html
Lufoin
http://lufoinregister.angelfire.com/B.htm
http://lufoinregister.angelfire.com/G2.htm
Project Hessdalen
http://www.hessdalen.org/index_e.shtml
SETI Search for Extra-Terrestrial Intelligence.

https://www.seti.org/
Some Inspiration (publications)
http://someinspiration.com/steps-ebook/
Trappist-1
http://www.trappist.one/
UFO organisations
BUFORA https://www.bufora.org.uk/
CNES http://www.cnes-geipan.fr/
CUFOS https://www.cufos.org/
DISCLOSURE PROJECT http://www.disclosureproject.org/
MUFON http://www.mufon.com/
MUFON-CES https://www.mufon-dsr.com/
NUFORC http://www.nuforc.org/
RUFORS http://www.rufors.ru/
SUFOI http://www.sufoi.dk/
UFO-Datenbank.DE https://www.ufo-datenbank.de/
UFO Sweden http://www.ufo.se/
Wikipedia
https://en.wikipedia.org/wiki/Abiogenesis#The_deep_sea_vent_theory
https://en.wikipedia.org/wiki/Ant
https://en.wikipedia.org/wiki/Cell_nucleus
https://en.wikipedia.org/wiki/List_of_potentially_habitable_exoplanets
https://en.wikipedia.org/wiki/Lost_City_Hydrothermal_Field
https://en.wikipedia.org/wiki/Symbiogenesis
Wonderpedia
http://www.wonderpediamagazine.co.uk/mind-body/why-our-body-temperature-37-degrees-c
YouTube
https://www.youtube.com/watch?v=wsEd_b1C8DY

Index

About the author

 Francis O'Neill writes on spiritual matters, mind body & spirit, and astrology.

As well as being a practicing astrologer, he has spent a good deal of his life in field archaeology, supervising rescue excavations on Roman and prehistoric sites, and also as a qualified lecturer in adult education.

He lives in the Cotswolds (UK) with his partner, the composer/musician, Annie Locke.

His book, *Love's Story of Why We Are Here* is a natural follow-on companion of his earlier work, *Life and Death: Making Sense of It*.

The Life and Death book investigates life and the afterlife through human evolution; the paranormal; near-death experience; past lives of children; past life regression; karma and reincarnation. With the aid of mediums and NDE studies and reporting, the book also provides evidence for the afterlife, and pays a visit to the Other Side. It is available in paperback and ebook formats. The book can be purchased through bookshops, online stores and also from the publisher website.

Find out more...

You can find out more about the writer's books, and his spiritual interests, by visiting the publisher website, SomeInspiration.com, where, like other online, and bricks & mortar stores, you can also buy a copy of these titles.

Reminder to leave a review ☺

If you have found this book of interest and helpful, please leave a review at wherever you got it from. It could encourage others to read it, and it will certainly encourage the author to keep writing. All reviews are read and very much appreciated. Thank you for your support.

Printed in Great Britain
by Amazon

17122584R00098